SCIENCE AND THE CREATIVE SPIRIT

Essays on

Humanistic Aspects of Science

by

KARL W. DEUTSCH

F. E. L. PRIESTLEY

HARCOURT BROWN

DAVID HAWKINS

Edited by

HARCOURT BROWN

for the American Council of Learned Societies

UNIVERSITY OF TORONTO PRESS

1958

Reprinted in paperback 2014

London: Oxford University Press

SCHOLARLY REPRINT SERIES
ISBN 978-0-8020-7012-8 (cloth)
ISBN 978-1-4426-5133-8 (paper)
LC 58-1232

The editor and author wish to acknowledge the permissions granted by the following for the use of copyright material in Essay III:

Victor Gollancz, Ltd., for the lines from Hugh MacDiarmid's *Stony Limits* quoted on page 86.

The Oxford University Press and Christopher Fry for the lines from *Venus Observed* quoted on page 86.

J. B. Lippincott Company for the lines from Alfred Noyes' *Two Worlds for Memory* quoted on page 87.

The Literary Trustees of Walter de la Mare and Faber & Faber Ltd. for the poem "The Moth" quoted on page 88.

SCIENCE
AND THE CREATIVE SPIRIT

Foreword

In February of 1950, nine scholars, chosen from the sciences, social sciences, and humanities, were brought to the offices of the American Council of Learned Societies, in Washington, to join with representatives of the staff in an informal conference on "Relationships between Science and the Humanities." The plan for the meeting had been developed by Charles E. Odegaard, then Executive Director of the Council, in order to obtain suggestions towards a continuing program. As had been expected, the conference contemplated a broad variety of topics and questions, including the influences of science on society and of society on science, and the interdisciplinary relationships and potentialities of the study of the history and philosophy of science.

The principal result was the appointment, in the summer of 1950, of a Committee on the Humanistic Aspects of Science, under the chairmanship of Harcourt Brown. The Committee was chosen to consist of "individuals who are not practising or professional scientists but who have been interested in scientific investigation as a human and social phenomenon and who, working from humane studies, are concerned with developing a better understanding of the role and functioning of science in human history."

After some preliminary bibliographical studies the Committee elected to prepare through its own membership the present collection of essays. That this volume comprises a collection of essays should, perhaps, be repeated: the authors chose their subjects and developed them according to their individual views and convictions. On the other hand, each essay is in a most real sense the work of the entire Committee. Each paper was repeatedly subjected to critical review of its style and content, both in meetings of the group and by correspondence, and revised accordingly. The members of the Committee who did not contribute essays were participants, not bystanders. The Committee held its final meeting in the spring of 1954, although work on the text continued for a long time afterwards.

The American Council of Learned Societies acknowledges gratefully various kinds of contribution to the program of the Committee. Several institutions provided congenial quarters, and in some cases added hospitality, for one or more meetings: Institute of the History of Medicine, The Johns Hopkins University; the John Hay Library, Brown University (which provided library space and service also); Washington Square Center, New York University; and Hart House, the University of Toronto.

The circumstances under which this volume is published offer a happy instance of international scholarly co-operation. Not only did the University of Toronto Press cheerfully accept the text and provide support from its Publications Fund, but the Humanities Research Council of Canada also contributed towards the cost of publication.

Members of the Committee on the Humanistic Aspects of Science who saw the essays through to completion were: Harcourt Brown (French language and literature), Brown University (*Chairman*); Bernard Barber (Sociology), Barnard College (*Secretary*); Karl W. Deutsch (History and political science), Massachusetts Institute of Technology; David Hawkins (Philosophy), University of Colorado; F. E. L. Priestley (English), University of Toronto; Richard H. Shryock (History of medicine), The Johns Hopkins University; William D. Stahlman (History of science), Massachusetts Institute of Technology; and D. H. Daugherty, ACLS staff liaison. Henry Guerlac (History of science), Cornell University, and Victor F. Weisskopf (Physics), Massachusetts Institute of Technology, were also associated with the Committee at one time or another.

D. H. DAUGHERTY

American Council of Learned Societies,
Washington, D.C.

Contents

Why This Book Was Written

An Introduction by
HARCOURT BROWN

A student of the humanities may be pardoned for hesitating on the threshold of this book. He can neither escape the problem of values nor avoid the inevitable accusation of partisanship; the literature of the subject is already shot through with emotional overtones and sentimental associations. The intelligible content of such words as humanist and humanistic, scientist and scientific, is not neutralized with the passage of time; and the invention and use of phrases suggesting a reconciliation of the opposites, *scientific humanist, humanistic sciences,* do not quite silence the uneasy thought that we are here not dealing in truths of fact or even perhaps of theory, but rather entering an ambiguous world, where ambivalent and subjective appreciations ultimately rule.

Thus, while some scientists deem it important to describe their quest of theory as a humanity, suggesting its imaginative and creative aspects, many humanists have sought to imitate the scientists, reworking their disciplines to meet scientific requirements of objectivity and cumulativeness. The justification or refutation of these contrary claims demands much critical analysis, some of which is at least begun in these pages; the task is hindered by the absence of a basis of mutual understanding, from which we might begin to revise the methods and objectives of either field to meet the requirements of the adopted appellation. In such an area as linguistics there was systematic observation and orderly classification long before it was claimed that the study stood on an equal footing with older sciences, such as physics or biology. The new adjective "scientific" expressed a desire for approbation—perhaps even self-approbation—rather than a radical change or even an extension of the scope and significance of an established activity.

Indeed, it would seem that in each great division of learning there remains a specific residue that is incommensurable and unique, a residue which it is part of the intention of the present writers to characterize. Science may be equated with knowledge, as one of our chapters suggests, but there will inevitably remain an important fragment of mankind for whom the mathematical way of thinking is incomprehensible, for whom accordingly an impressive part of science will not be available. It is a common failing among professionals, scientists and humanists alike, not to allow for the blind spots in those once styled the "Philistines," but to expect full understanding from the layman; yet a moment's reflection brings to mind those with whom one cannot profitably discuss the fine arts, or perhaps the theatre or poetry, or social problems, or a branch of science. Early training has of course much to do with the variations of mankind; but there are obscure differences between human beings which suggest that we cannot all be scientists or humanists, just as surely as we cannot all be virtuoso violinists or even sing our national anthems with true pitch and self-confidence. There is much to be said for defining our terms so that communication is effectively attained, and distinctions recognizing real and important differences are not blurred.

I

Old and respectable tradition associates the humanities with the study of the record mankind has set down in books, the information available concerning man himself, his taste and morals, his manners and thought, his social relations and his beliefs, his arts of peace and war, his relations with the world within himself as well as with nature and the elements. From the wide field of interests represented in the library of the Renaissance humanist have been derived many specialized modern disciplines: history, geography, natural philosophy as distinct from divine or supernatural, political and social science, aesthetics, psychology, each of them once undistinguished and largely undiscriminated parts of the study of humane letters. Recent developments, subdividing still further the remnant of what a fully equipped scholar once took for his province, have increased the number of intellectual splinter groups: musicology, numismatics, the history of art in general or of the arts in particular, bibliography, linguistics, have not so much diversified the qualifications of the one-time humanist as created so many new varieties of specialist. What has happened in the tremendously diversified field of modern science has been taking place, though not perhaps with quite so much necessity, among the so-called humanities, with the result that there has been nearly an eclipse of the large spirit which once admitted any kind of book, provided only that it dealt in a broad way with man and some of his interests, to the study and library of the man of taste and intelligence. If the central area of the humanities be the study of books, the arts that go to make them, and the kind of knowledge they contain, one may read a description of what goes on in this field in the Report of the Committee on Research Activities of the Modern Language Association of America in *PMLA*, October, 1952; it is discouraging that even that small nucleus is not yet irreducible.

Indeed, the specializing tendency is so strong that it seems improbable that we shall see a restoration of the old full richness of the humanities—the spirit that animated a Montaigne or a Rabelais, a Shakespeare or a Milton, as well as the schools and *salons* and academies of their age—except in sporadic and inconclusive efforts. Each modern study that is set up is instructive in this respect; scholars from different established disciplines converge to explore a new pathway, leading to an area bounding different known fields of thought, in the hope of broadening the outlook of their contemporaries and posterity, only to end by training new specialists in a new and restricted compartment of knowledge. One may instance the development of such a discipline as the history of science; although at one time it was a common hope that this new field would result in the production of more tolerant and broadly educated humanists as well as more humanized engineers and research scientists, the principal result seems to be another specialized profession, with its own content, its own methods, its own more or less independent sources of vitality. It does not seem to be transforming the outlook of the scientist, or making a serious impact on the training and qualifications of the humanist.

Divided, these disciplines have advanced and become strong in their own special domains. In varying degrees they still possess certain qualities characteristic of the tradition of humanism; some of them express an interest in objects and ideas for their own sake, to be contemplated without reference to their immediate utility or their ulterior purposes: objects which reward those who look at them with a special kind of pleasure, with varying degrees of inner satisfaction and delight, or ideas which lead to nothing beyond a more satisfactory understanding of man and his capacities, or of the world and the universe around him. The mental act, the inner power that brings forth these objects of contemplation may not be in fact very different from the intellectual activity that is the source of science; however, we can recognize two or three characteristics more emphatically present in humanistic contemplation than in scientific analysis. We have suggested that the humanist is not concerned with use, with tools, with the ulterior purposes of the objects of his study. It could be added that he is not so much interested in power, in what Francis Bacon described as the purpose of Salomon's House, "the knowledge of causes and secret motions of things, and the enlarging of the bounds of human empire, to the effecting of all things possible," as he is concerned with things for their own sake, for their value as expressions of the human spirit. Humanistic interest respects the creative powers of man without seeking to submerge the individual object in the service of an abstraction like society or science; it does not ask, with the laboratory scientist, what can be done with a statement or an artifact to lead at once to something new. The object of the humanist's attention is the creator, the individual man, or man taken as a moral type. From the beginning of what we may describe as humanism, through such texts as Plato's *Ion*, the *Symposium*, the *Phaedrus*, and others, the scholar has sought an ever more exact account of the relation of the artifact with the whole human being who produced it.

Perhaps, then, in an effort to give the word a useful modern context, we may describe the humanist as one whose object of study is the creative imagination of man, in its efforts to produce objects and material effects—poetry and music, sculpture and painting and architecture, bridges and palaces alike—all the arts in short, whatever works of a representative or structural nature as may possess a high degree of aesthetic or moral purpose or content, proportionally greater than their utility or factual communication. The humanist devotes his efforts to the understanding and explanation of man—man as a creator interested in creating for its own sake, without regard for ulterior satisfactions differing in kind.

So it is clear that when we try to apply the adjective *humanistic* to science or to any aspect of science, we find ourselves caught in a tangle of paradox. In the first place, even in speaking of its "humanistic aspects," we must work from a view of science that the scientist will neither reject outright nor refuse to admit as a recognizable appearance of his subject. Normally susceptible to comments on his work deriving from persons outside of his tradition, he may legitimately question the basis of such comments and of the inquiry from which they stem. Yet it is also apparent that our objective here is to review, as candidly as a small volume will permit, a complicated field in which some essential aspects of science are to be presented as they appear to non-scientists—friendly, we hope, but still non-scientist—scholars who are responsible for their views to the ethical standards of scholarship and to their own consciences. It follows that what is to be presented here is intended neither as a defense of nor as an attack on science. Just as we regard the word "science" as having a precise meaning which must be respected, so we have attempted to keep a degree of precision in the word "humanism," with its derived forms, *humanist*, and *humanistic*. Little is gained by allowing these words to lose their virtue to anyone who regards their undoubted emotional values as something to be conquered or pre-empted, and more significant than their reference to arduous disciplines, to be studied and understood.

Furthermore, if we are to speak of an "aspect" of science, it follows that we must be objective, perhaps in several different ways, about science. An aspect is an appearance, a perspective seen and appreciated from without; we shall have to define and describe science as a going concern with its own rules and purposes, but in evaluating its aesthetic and creative aspects we shall, as humanists, in the nature of things have to see it from the street, as something present for a long time on our intellectual highways, which has had a determining influence on the shape and function of nearby buildings, and which a neighbor is entitled to think about in various terms and relationships. After three centuries of clamoring for recognition, no scientist can protest if neighbors seek a full-length view.

It should perhaps be emphasized that what we offer here is rather an evaluation than a defense or eulogy of science. It will be apparent that none of us has much inclination to attack it; we hope it will be equally clear that those who have contributed to this volume do not feel that criteria external to science are irrelevant. What science has done is significant in its own right, in comparison with the fine arts, with the long calendar of the world's literature, with the attainments of scholarly humanism, and with the humanist preoccupations of the philosopher. We believe therefore that the true humanism of today demands that science be included in the full account of mankind, that its contribution to man's life in freedom, to the creative and non-utilitarian part of man, be set down in order.

There is still another implication in this book; it is that science may be treated as a unit, as something that can be grasped and evaluated in its diversity as a discipline with unique and typical goals, possessing a characteristic method or methods derived from its objectives and the conditions in which it must operate. The paradox mentioned above consists in the difficulty one faces in discussing the freely creative aspect of an activity that has largely sprung from the useful arts and which in the end returns to create industry. We have not felt entirely free to discuss science apart from its applications; different as theory and practice may be when seen from within the fabric of science, from without it is extremely difficult to decide where one begins or the other ends. Modern science, in its most advanced aspects, employs large and very complicated machinery, in turn depending on precision tools and techniques, which are finally based on elaborate theoretical structures. Parallel with this relationship, a fully equipped laboratory working at capacity depends on a great range of factories refining metals and other materials, producing electronic apparatus, high precision lenses, and all the rest, to standards which can be maintained only by an industrial structure organized in great detail and operating on a very large scale. The author of one of the following papers has remarked on the increasing degree of coupling between science and its social environment in modern times, and has pointed out that this social change "means, in one way or another, the disappearance of that degree of autonomy which (scientific) research has mostly enjoyed hitherto," and he suggests further that, in the process, the ethics of science, which so far has been in large part a celebration and defense of this autonomy, will have to become an ethics in which science seeks the freedom of society rather than freedom from society.

Perhaps therefore the layman is justified, although not excused, in seeing science in terms of its manipulative power; he sees behind the services it offers, and the hardware it consumes, a certain structure of ideas about life and the world, and also a program of operations involving a knowledge of techniques and processes which may be hidden for reasons of profit or security. While there may be a real distinction between the theory and the technology, to the layman, as to many even among the scientists, they look much like two façades of one building, each essential to the other. To change the figure, each feeds the other, differently, but in reality. In any case, in recent times, bitterly as the pure scientist may decry his bondage to public and private interests, he has had no impulse to bite the feeding hand. Indeed, much of the talk about the essential nature of research in pure science has been devoted to the argument that the development of basic science is really after all the most efficient way to an improved technology.

II

At the heart of any discussion of the humanities lies the area of imaginative creation, the domain of the muses—the arts of painting and sculpture and their derivatives, music, the dance, drama, the poetic forms of literature, the epic vein in history, prophecy: all forms in which inspiration joins with memory, Zeus with Mnemosyne, to elevate and glorify the destiny of man. Ideally, perhaps, the arts are free of objective reference and didactic intent, but frequently—and this is a source of confusion—in unsophisticated times they have been used to teach or to represent ideas and objects important to the artist or the author, on behalf of the community, in the desire to share and perpetuate the sentiment or evaluation embodied in the work of art. In this way the arts may offer the modern student information of an historical nature, in epic or legend, or data concerning the religious sense and the theology or ethics of the author and his community; in many cases we can learn from the work of art something of the science and the social customs of the community as well. But content in these terms does not account for the whole work of art, which is more complex. One cannot abstract any component, scientific, religious, ethical, or historical, and say that what is left is pure art—poetry, painting, music, or the like.

It follows that a work of art can be translated from one idiom to another only very approximately, and with great difficulty. Painting remains painting, and refuses to be described in words or music, which equally cannot be rendered in color or in verse. A poem by Mallarmé may suggest music to Debussy, which in turn inspires a ballet for Nijinsky; but the three forms, related by historical association and vaguely parallel, are distinct and incommensurable. A hard core remains in each, untouched by the second artist's hand and skill. Science, on the other hand, has a universal quality, not only in the sense that it is impersonal as between one human being and another, but also that it may be rendered in any civilized language; furthermore, much of it may be described in more than one set of terms. Thus the motion and path of a falling body may be described in algebraic formulae, or on a graph, or in terms of a verbal description, which may in turn be translated without significant loss of precision into a hundred languages. A poem of Baudelaire, an *étude* by Chopin, a painting by Cézanne, a sculpture by Michelangelo, defy such conversion. *Traduttore traditore.*

If we accept the creative imagination of the artist as the heart and center of what we refer to as the humanities, we may agree that historical and critical studies of this imagination and its works comprise the first level of humanistic scholarship. Here we will find the history and description of the arts, critical evaluation, theoretical analysis, and social and psychological explanation; as these forms of study become clarified and self-conscious, specialization takes place, objectivity is deliberately sought, and science may be said to begin. But we can still distinguish the critic who allows for the individuality of the artist and his audience from the scientific scholar who seeks to classify the reactions involved, who is content to describe Molière's Alceste as a psychopath rather than as a normal, though vividly extreme, product of the rationalist psychology of his age and country. It is perfectly admissible that an art historian or a critic, even one with some theoretic interests, should share in the quest for individual expression and in his analysis of artistic activity refuse to seek firm categories as a basis for the kind of general laws and uniformities which are the beginnings of science.

Distinct from the analysis of form, and yet equally fruitful from either of the points of view just described, is the analysis of intellectual content. The earliest forms of art possess a representational content, referring to the natural world, the passage of time, the recurrence of important events, the shape and color and behavior of familiar objects and persons. Later, the interest in particulars develops into a sense of things in general, their common aspects and qualities, seen or experienced in themselves apart from the observer who becomes a mere recorder of quantities and dimensions. This could be considered the beginning of science: when the artist becomes aware of alternative ways of expression, of the element of truth that can be put in different ways by means of suitably devised techniques, when he learns that a formulation may be true in many mouths and in many different contexts, and when he learns that in saying it thus it is released from its "humanity," from the individual variable, partly personal, partly a function of his tools and his chosen medium, which he comes to regard as an essential component of any work of which he can be proud. But the way of the artist, with his individual vision, is separate from, and perhaps incompatible with, the way of the scientist, who has much less patience with or interest in irreducible idiosyncrasies.

While no doubt every articulate man has something of each spirit in him, it is still possible to distinguish between the humanist as self-communicating and variable and the scientist as objective and impersonal, and to apply this distinction to the spectrum of intellectual activities which lie between the extremes of the sciences and the arts. A middle ground would be occupied by such disciplines as History and Philosophy, for in them the human need for emotional and aesthetic satisfaction is present along with objectivity and impersonality. If any particular history is to be widely read, it must in some measure be the history of the reader and a group to which he can at least imagine himself to belong. Likewise, many readers of philosophy do not clearly distinguish between rational and impersonal intellectual analysis and emotional release and consolation. When the historian explains and organizes his material in terms of numerical formulation, using statistics and economic factors to account for events in historical sequence, we may agree that he is approaching the scientific outlook. On the other hand, if he is aware of a personal attitude towards his subject-matter, if he attempts to write with elegance and charm, employing stylistic devices and introducing elements of an impressionistic nature for purposes of filling out a picture that extends beyond his controlled factual data, if he finds it difficult to avoid dramatic aspects of his story, to escape the urge to treat men as dynamically independent entities rather than as units in a quantitatively homogeneous mass, if he finds an imaginative use for the arts, for the unpredictable creations of craftsmen and the exhortations of maverick prophets, then we may say his work approximates that of the humanities. The diversity of philosophers resembles the diversity of the historians. As a servant of the freedom of man, in ethics, in the arts, in analyzing man's reflections on life itself or its great events, on the ends of conduct or existence, the philosopher is a true humanist, in that he uses intuition and imagination to produce results beyond those justified by operationally measurable aspects of things. But in the pursuit of symbolic logic, in realms where reference must be made to objective cumulative knowledge, to repeatable experiments, to an impersonal structure of thought, philosophy is not one of the humanities as the word is used here.

We may summarize the situation as we see it by suggesting that the humanities as we shall discuss them include: the creative arts, largely interpreted; the various critical and historical disciplines that describe, explain, elaborate, or theorize on the basis of these arts; discussions of man in respect of his activity or interest in the arts, or in the critical, historical, or theoretical aspects of the arts; and the theory of such discussions, the relevant aspects of knowledge, of history, of criticism, aesthetic experience, imaginative creation.

If we attempt to look at science from this point of view, to discover what might be described as the "humanistic aspects of science," we find ourselves in some difficulty immediately. If science has an aspect which may be called humanistic, it is clear that it is in some degree accidental and unintentional. From this point of view, a scientist would have no very marked desire to see his work described as an "individual variable"; nor would he like to have his work grasped intuitively to the exclusion of rational understanding. He is not interested, as the artist has usually been, in the creation of a final and eternal form in which experience is seized and perpetuated. The modern sense of the "endless frontier," of the "unending quest," to quote key phrases from the titles of recent books, has largely broken down the view that the fabric of science has the same kind of permanence that poets have sought ever since the *monumentum aeri perennius* erected by Horace.

In fact, the scientist's work is tentative, a brick in the cathedral of physics, as one author describes it, a fragment of the total structure; and we may add that, like any brick in any other structure in constant use, it may be withdrawn and replaced by another the moment its inadequacy becomes apparent. It is always possible to imagine that a better formula may exist, a more precise observation or more accurate description, a more exact statement of a law. On the other hand, the frescoes of the Sistine Chapel, the Fifth Symphony, the *Divine Comedy*, have a durability and permanence independent of serviceability in a structure, or of the support or corroborative value they may lend to some other product of the arts. Since works of art do not easily and usefully lend themselves to comparison with one another, it is clear that they cannot usefully be compared with works of science, which have an entirely different function and purpose. The artist's interest in his work is not self-effacing; as his problem is unique with him, he is the only man who has a chance of solving it, and solve it he will if time and skill permit. He strives for durable form, a structure which does not depend on other forms for its relevancy and significance.

Tout passe.—L'art robuste
Seul a l'éternité,
Le Buste
Survit à la Cité.

But if science is not inherently humanistic in the sense we have suggested, it has at any rate important humanistic implications. As Karl Deutsch will point out in greater detail, science has relationships, points of view, methods, and has had results, significant for the arts, for religion and ethics, for broad historical perspectives. In innumerable special areas, science has touched man in different ways, not all by any means merely useful to health and comfort. Yet the production of standardized articles in vast numbers has not inevitably led to the sacrifice of individuality; the sheer complexity of the industrial structure has made immense variety possible in color and arrangement, in complete freedom of taste. Textiles, for instance, offer today a range of qualitative differences many times larger than a century ago; curtains of half a dozen synthetics, woven alone or in combination with other traditional fibers, tinted or dyed or printed in several different ways, in colors beyond enumeration, may now be employed with a great variety of materials old and new, plastics, rare or familiar woods, metals, pure or alloys, ceramics and enamels, to provide an enormously enlarged spectrum from which the individual may choose a background for the idiosyncrasies of his personality. This condition is neither new nor accidental; it stems from the constant interplay of knowledge and taste, of science and art, which became notable in the Renaissance with the deliberate introduction of silk production from the Orient to Europe. The trend, which illustrates admirably the narrow boundary that separates the history of the arts from that of science and technology, accelerates with the discovery of chemical equivalents of rubber and silk in the late nineteenth century.

Thus there is no reason to believe that the background of the individual human being is likely to be reduced to a limited set of standard patterns as long as the normal development of science and technology is not hindered by the actions and negligences of man himself. On the contrary, an increasing range of materials and forms tends to develop the capacity and necessity for choice, as the number of criteria employed in selection increase, and thus to enlarge certain aspects of the intelligence of the consumer. It would be interesting to have a documented study of the operation of this influence in western society over the last three or four centuries.

Many observers have remarked, however, that the gain in opportunity and discrimination has been accompanied by a significant loss. The very immensity of choice possible in a scientifically-equipped culture breaks, or at least dilutes, the aesthetic and artistic tradition, and changes the quality of the community in comparison with others less exposed to such influences. We need hardly dwell on the impact of new modes of printing on the patterns of the necktie, on changes in shape and function imposed by the limitations and possibilities of new plastics, formed by new methods, or on the changes in political and social habits attributable to new methods of mass communication. Closer to the heart of the matter is the effect of a new technology on language and habits of thought and feeling. There are many poets whose work recalls comments made by Bertrand Russell on the value of long usage, on words "coloured by embedded emotion," which lose their charge and to that extent their poetic value, as the objects to which they refer pass from common knowledge, and their content becomes a matter of antiquarian information rather than symbols of a way of life. Once one could write about a horse in a way that is ridiculous when applied to a jeep or a tank; the "open road" is a phrase without meaning on the highways of today.

It was with such developments in mind that Russell could suggest in the 1920's that "it is possible in poetry to write a letter, but difficult to speak over the telephone; it is possible to listen to Lydian airs, but not to the radio"; and he added with much justification that the matter-of-factness which science has brought to the language of the telephone and telegram has destroyed many of the attitudes and poetic formulae which depended on the size of the world and on the consequent slowness and difficulty of communication. Thule photographed and charted and set down a few short hours from Rome and Athens and New York ceases to merit Gretchen's song; its life is a pale reflection and transposition of our own, even in Arctic wastes under wintry skies. Russell concludes that "the aesthetic effects of science have thus been on the whole unfortunate, not, I think, owing to any essential quality of science, but owing to the rapidly changing environment in which modern man lives."

Russell observed what had happened to the heritage of the past, but he did not then perceive that the result had been not so much a destruction of poetry as a transformation of it. When he wrote, new movements were already astir in all the arts, in poetry as in music and painting and the theater, in which the effect of the new tempo and mode of life were clearly visible. Sentiment was giving place to irony; Noel Coward and Aldous Huxley, T. S. Eliot and André Gide were finding the source of a new aesthetic in the world emerging from the First World War—an aesthetic which is the result of the development of a new attitude towards the world in which man finds himself. In a few centuries the influence of the sciences has turned many men from contemplation to manipulation, from love of objects for themselves existing in a particular context, with rights and functions with which they would not interfere, to a love of power for its own sake, or for private and personal advantage. As Toynbee points out, applied science has advanced beyond either political control or religious sensitivity. There are modern publicists who give expression to this attitude; in their perspective, the universe, the natural world, even man himself, and his most abstract creations, his music, his mathematics, nuclear physics, are but tools in the hands of an engineer who respects nothing but his blueprint.

The result has been a widening gap between science and the humanities; science is influential among modern men of letters, but it has not persuaded them. Russell draws a contrast between science as he finds it among the Greeks and the science of the twentieth century. It was in the beginning a poetic, personal love for the world as it appeared to the senses; it was grasped in a kind of beatific vision, a mysticism which sought union with the world by means of a greater understanding of scientific law. Greek science was an expression of a love for the world felt "almost like a madness in the blood." In place of the sensory knowledge we once had, of color and sound, light and shade, tone and silence, the ultimate realities of the artist, we are now asked to accept path and number, charge and mass, units of energy which are imperceptible to our senses, phantoms of the mind, visible only as shown by a pointer on a dial or a curve on a graph. Transformed, along with the principles by which society operates and develops, into statistical probabilities, acting impersonally, with neither respect nor malice towards man, the laws of science became tools for the execution of projects involving control over nature and over the lives and conduct of masses of men. It is no wonder that in the process the sense of the brotherhood of man is lost, as human values tend to disappear in the calculation of forces, inertias, energies, resistances, and the like. Russell suggests a fear common to all those who knew the 1920's when he writes that "only in so far as we renounce the world as its lover can we conquer it as its technician."

The views and the reactions of those who could observe the materialistic twenties of this century deserve the serious attention of the humanist of today, for they represent a spirit that underlies much of the anti-scientific revolt of modern literature and art, and with them of humanistic scholarship, not only in America, but also in England, France, and elsewhere. Russell's view that the typical modern man is a being devoid of culture, contemptuous of the past, self-confident, ruthless, mechanically minded, deserves constant and systematic checking against the records of our time. The question must be asked whether there is not great risk that the scientist is condemned by his science to a devastating parochialism of outlook, to one kind of thought, one kind of criteria, one kind of end. For all his scepticism about words and phrases, his neat slogans—*Nullius in Verba*—has he in his mind the leaven of doubt about himself and his powers that makes a man and not a devil? The *Bulletin of the Atomic Scientists* suggests the possibility that science can be a broad and human enterprise with many humanistic aspects. But there are still scientists who never leave the realm of science long enough or far enough to see it in its historical background, its social context, or its human inadequacies.

And it is precisely as it acquires this capacity to see itself in its context that science may become a complete partner in modern culture, demanding the attention of the humanist who tries to see modern life fully and fairly. It is not enough to train still more scientists to communicate techniques, to teach the facts and methods and inculcate the attitudes of the laboratory, indispensable as these may be. There must be—as indeed there often is—a responsible attitude on the part of scientists, responsible to the whole human scheme of things, to humanity. The scientifically trained technicians who will in time dominate our affairs through sheer capacity to manipulate immense resources of materials and energy by the use of vast electronic computers must know and respect the origin of their authority, the nature of the human tradition, the duty they have to keep life tolerable for the governed. "Scientific technique," says Russell, "must no longer be allowed to form the whole culture of the holders of power. . . . Science, having delivered man from bondage to nature, can proceed to deliver him from bondage to the slavish part of himself." In his *Brave New World* (1931), Aldous Huxley was not so sanguine, and in a foreword to that book dated 1946 he places the responsibility further back: "Unless we choose," he writes, "to decentralize and to use applied science, not as the end to which human beings are to be made the means, but as the means to producing a race of free individuals, we have only two alternatives to choose from: either a number of national, militarized totalitarianisms, having as their root the terror of the atomic bomb and as their consequence the destruction of civilization . . . or else one supranational totalitarianism, called into existence by social chaos . . . , and developing into the welfare-tyranny of Utopia. You pays your money and you takes your choice."

III

The four essays which follow have been written with a view to exploring the general area suggested above. While each is signed by one person, and should be regarded as the product of an individual approach to its own theme, the papers taken collectively show the influence of the Committee as a whole. Each has been debated at some length by the Committee, and most of the ideas and points of view subjected to sympathetic but sharp and searching critical discussion. If the various authors do not all say the same things, or even represent an entirely unified viewpoint, it is because none of us felt that we should force a compromise on what as individuals we deemed essential, or that we should seek an artificial common ground when our areas of particular competence and professional loyalty did not entirely coincide. At this point in our thinking it is desirable to seek variety, to explore the range rather than to attempt to concentrate. Many points of view, many angles, much breadth may suggest the richness of the area of our interests.

The members of the Committee, whether writing papers for this volume or not, desired therefore rather to open up an area for research and discussion than to set forth a final and considered opinion. Each statement should be thought of rather as a question to be discussed than as a dogma to be believed; in several directions we have been inclined to indicate possible findings rather than demonstrated results. Our hope is that the general bearing of the book will have more influence than its substance, in some parts at least of which the ice of conclusions lies thinly over waters of scholarship of uncertain depth.

The reason for this necessary warning lies of course in the fact that none of us are specialists in this particular field, which has so far been explored largely on an amateur basis. The members of the Committee teach philosophy, various branches of history, political science, sociology, or literature; their reading and teaching has brought them to take an interest in the impact of science on their respective fields, and the American Council of Learned Societies has enabled them to meet for discussion of these matters with some regularity over a period of about four years. In the course of these sessions or seminars, points of view were set forth and elaborated, criticized and debated, and in some rare cases accepted by the group as a whole. The general use of a simple assertive statement throughout these papers should deceive no one; it is a stylistic device which seemed more acceptable as we read each other's papers than a continued subjunctive mood, with accompanying *mays* and *mights*.

In spite of this disclaimer, it may be suggested that there is a possibility that some progress has been achieved here, that the collective work of the group has perceptibly extended the area of the humanities. We think we have outlined an aspect of our culture, developing over a considerable length of time, crossing national boundaries, and that we have brought together facts and findings from a number of hitherto unrelated fields of knowledge. Science has been an independent component of our culture for three centuries, roughly speaking; it has admittedly made a deep impression on literature, the arts, religion, morals, and philosophy, and it has made a serious difference to the temper of modern scholarship. All this is quite apart from its enormous influence on the material conditions of life, in terms of comfort, convenience, and improved methods of mass destruction. It seems to be time, therefore, to separate the history and sociology of science from what we have been inclined to leave under the phrase devised by Charles

Odegaard, who called this Committee together and assigned it its task, the "Humanistic Aspects of Science." There is no one word that would define our studies; one might suggest that they bear a little of the relation to science proper that musicology bears to music, not the direct study of music, nor yet its history, nor yet the sociology or philosophy of music, but rather the study of the traces left by music in non-audible form, written, printed, iconographical, and so on. So our range of interest inquires into the traces left by science in the arts, literature, man's thinking about himself, and into the reciprocal influence of many external factors on science itself.

Although we have as yet no useful name for our studies, still we feel that it is possible that reasonable consistency, a deliberate search for unity of outlook through reconciliation of the different traditions in which we have been trained, may in the long run produce a structure of ideas, a method, and a subject-matter sufficiently individualized to deserve recognition among the humanities. At present, there is no avoiding the impression that each chapter demands a degree of reorientation from the reader; there has been, however, considerable effort to meet the editor's requests for a mode of expression that would permit the humanist reader to perceive, under the obvious contradictions and inconsistencies of the freely expressed views of a group of individualists, a serious effort made by each to view the complexities of this subject with critical sympathy.

IV

The humanist recognizes the scientist more often than he does the science; he is aware of human limitations in public relations rather than of intellectual powers and theoretic capacities. He rejects the tendency to abstraction, the desire to treat the world in manipulative terms rather than for its power to delight. Rightly or wrongly, he sees the scientist as one who intervenes rather than admires, who transforms and adapts rather than accepts. That a scientist uses a natural force to overcome an equally natural object or obstacle is perhaps understood but it is not necessarily accepted by the humanist; what the non-scientist sees is the intervention of a technical device, with all its artificialities and ornamental deceptions, in place of, or in contrast with, natural forms and textures. Electric transmission lines, highways, dams, canals, bridges, railways, factories, and the accompanying housing developments, shopping centers, and so on, are perhaps inevitable accompaniments of industrial and urban civilization, but there are surely various opinions possible about what such improvements improve. In general, it may be suggested that the humanist has no real quarrel with science, so long as it remains analytical, observational, experimental, knowledge-seeking. He is disturbed by the applications of science, by a science that is not content to remain knowledge but persists in remodelling not only the physical universe but also the structure of knowledge itself in the interests of a more perfect formulation. The humanist is a relativist tolerant of incompatibilities, undisturbed by inconsistencies even in the structure of his own knowledge, aware on good authority that there are more things in heaven and earth than may be dreamed of in any philosophy. He lives—not infrequently by preference—in an atmosphere of imprecision and undefined impressions that the scientific mind could hardly tolerate. Normally he is not accustomed to drawing a sharp line between himself as observer and the object of his contemplation; the very richness of his private life, the wealth of his perceptions, are a result of his willingness to submit himself to a kind of mystic fusion with the object of his thought. While the scientist may do something of the same nature at the height of his creative moments, he does not, as do the artist and the creative critic, usually try to reproduce the very nature and quality of the unique moment of delight or insight.

From another point of view, it has been suggested that the humanist, as well as the more characteristic Christian, rejects science as an ultimate guide because science rejects this union of the individual intelligence with the object of its contemplation. Science insists on defined relationships, measurable proportions, precise discernment of parts and limits, and the exact estimation of changes in state and position, and thus cannot accept "absorption in the infinite" as a clear and distinct idea capable of definition in operational terms. There are many instances of this break in the continuity of knowledge. William Blake offers one term of the contrast, just as Voltaire offers the other. The poet's sense of the "fearful symmetry" of the Tiger, of the "intellectual thing" that is the fly, of the unknown Maker of the little lamb, protests against the scientific rationalism of a materialistic age expressed in arithmetical precision and exact propositions, such as those sought by Voltaire's giant from Sirius,

Micromégas, as he questions the secretary of the Saturnian Academy and the astronomers of the Lapland expedition.

Thus there is much to be gained by a review of the issues involved in this extensive and disputed boundary area. That there is antagonism between scientist and humanist is hardly to be gainsaid; that it arises from misunderstanding and conflicting viewpoints seems to be quite as true. Perhaps no effort to trace the line of demarcation is futile; there is a pressing necessity to discover the interest each has in this twilight zone, where the insight of the scientist meets the evaluation of the humanist, where the humanist meets the passion for precision and operational advance of the scientist. Each may profit by a review of what he does and does not stand for. Little is gained by a polite concealing of essential differences and distinctions. The effort to find or to create unity in knowledge is laudable only so long as it recognizes the extreme difficulty of the operation and the strength and importance of the different impulses involved. Humanist and scientist both seek knowledge, but in fact the knowledge each seeks is different in quality, in kind, in use and function, and one word scarcely contains their diversity.

Therefore, while this book results from an effort to reduce the area of disagreement between humanist and scientist, it does not seek that end by denying that conflict exists. It is precisely because it is as hard for a scientist to see the problems of the humanist apart from the humanist himself as it is for a humanist to grasp the issues and principles of science beyond the individual scientist, that we think this group of converging statements on the Humanistic Aspects of Science may serve a purpose.

I

Scientific and Humanistic Knowledge in the Growth of Civilization

by
KARL W. DEUTSCH

THE SCIENTIFIC AND THE HUMANISTIC FOCUS OF ATTENTION; PROBLEMS OF CUMULATIVE KNOWLEDGE AND OF SCALE; QUALITATIVE ASPECTS OF SCIENCE; RATIONALITY AND INTUITION IN THE ORGANIZATION OF EXPERIENCE; PROBLEMS OF VALUES; SCIENCE AND CIVILIZATION AS MUTUAL RESOURCES AND CONSTRAINTS; THE JOINT ENTERPRISE OF GROWTH

I. THE SCIENTIFIC AND THE HUMANISTIC FOCUS OF ATTENTION

Lovers of letters and of life have looked upon the world but have found it hard to understand; scientists have tried to understand the universe, but have split it into fragments; how are these two viewpoints related to each other?

What is the center of our attention when we are engaged in scientific work, and what is the chief object of our attention when we are looking upon life from the viewpoint of the humanities? Before we can try to answer this question, it may be useful to indicate the meaning which will be given to the words "science" or "scientific knowledge," on the one hand, and the words "humanities" or "humanistic knowledge," on the other, in this discussion.

As the distinguishing mark of the humanities, we shall take their emphasis on the "whole man," on the interrelation of all his ways of thinking, feeling, and acting; and on the impossibility of subordinating the fullness and unity of his life to any single purpose, or to any single point of view. Even when the poet, the painter, or the composer is trying to convey only the particular mood of a particular moment, his concern is with the integrity and wholeness of this mood, however short in time. Art is thus essentially "presentational" in Susanne Langer's sense of the term;[1] it is based on the simultaneous perception and appreciation of many aspects of individuals or situations, and on the making of an intellectual and emotional response to some of the new forms presented by their combination.

Science, on the contrary, sets out to isolate a small number of such aspects, as well as to isolate events or situations from their contexts. Some of these isolated aspects or events are then selected and investigated by the scientist with utmost thoroughness, and a few of their implications are pursued with single-minded persistence. From this single-minded investigation of isolated aspects of reality, specific propositions and predictions are derived; these then can be verified, directly or indirectly, by the observation of other isolated aspects of reality, provided that in the case of each observation or experiment, all other circumstances can be safely neglected or controlled.

Creative art, and generally humanistic knowledge, cannot be

adequately tested by verifying any particular predictions that could be derived from it. Rather, the work of art or the literary portrait of a personality rings true if we can perceive it as a whole, if it fits some configuration of memories preserved from the past, or if it helps us to create within our mind a new and valuable configuration of symbols or ideas for the future.[2]

Scientific and humanistic knowledge are thus two different sides of one and the same process of human thinking. Knowledge is a broader concept than mere data suitable for tabulating. It includes such data and quantitative formulae. But it also includes understanding, "knowing" what to expect, how to act, and even how to compose or manage one's own mind. It includes intuitive knowledge, or what Pascal called the "esprit de finesse," as well as the "tragic knowledge" of such philosophers as Karl Jaspers.

Within this wide realm of what men know, scientific and humanistic knowledge are as inseparable as the intellectual processes of synthesis and analysis, or as the presentational and the discursive views of symbols and language. The scientific mind, like Plato's, must follow its argument "wherever it may lead," provided only that it is supported by operations of verification within a term of self-imposed limits. The humanistic mind, on the other hand, will refuse to submit to such one-sidedness. Science will be most concerned with those aspects of reality that can be organized in terms of repetitive and public facts; and by "fact" the scientist means either events which can be directly reproduced in all their relevant aspects, or events which can be inferred indirectly from the evidence of the reproducible traces which they are believed to have left. Since the humanist, on the contrary, is interested in the whole situation, and not necessarily only in its repetitive features, he is apt to be interested in its unique aspects, as well as in the private aspects of perception or experience by individuals.

The sciences and the humanities are thus situated along a single more or less continuous spectrum of human activities. No science can be so abstract, so single-minded, so preoccupied with repetitive facts or repetitive symbolic operations that problems of simultaneity and relative uniqueness, or of individual perception and correlative recognition, may be excluded from it altogether. And no form of humanistic knowledge, on the other hand, is entirely free from the problem of repetitive consistency in its own views of patterns and symbols, or from the problem of its relationship to the world of repetitive facts.

Science thus forms one essential part of human activity, taking its place beside other parts no less essential. It forms a part of the growing up process of every child, as he remembers and orders his experience, and as he begins to project inferences from it into the future; and it becomes the main vocation of some adults. On the social level, science is found in the activities of every ordinary human group which abstracts data from its experiences and uses them to guide its future behavior.

At a more complex stage, science becomes a major social enterprise, carried on by research organizations, laboratories, universities, and other institutions in every modern industrial economy or culture. Science finally represents a link bridging long periods of time, both between different memories and experiences of an individual, and even more characteristically between widely separated generations of the human race.

The Problem of "Accumulative" Knowledge

With regard to this last "time-binding" function, science has sometimes been called "accumulative knowledge," notably by James B. Conant, and the suggestion has been made that the arts and the humanities are not cumulative in this sense.[3] To what extent is this alleged distinction between "accumulative" and "non-cumulative" kinds of knowledge justified?

"Accumulative sciences," according to Mr. Conant, are those natural sciences in which conspicuous and undoubted advances have been recorded over the recent centuries, in such a manner that we may presume that even the early pioneers of those sciences, such as Galileo or Harvey, could they come back to life today, would agree that substantial advances had indeed been made. "Non-accumulative knowledge," according to the same reasoning, would essentially consist of those fields of art and philosophy in which no such general agreement about conspicuous and undoubted advances would exist; Rembrandt, Dante, Spinoza, Locke, or Kant, all would not necessarily agree, according to Mr. Conant, that significant advances had been made in their respective fields since their own day. Moreover, no clear agreement could be found among ourselves, if we were to discuss the question of progress or retrogression in the arts or in philosophy today, nor would there be any "agreement between the majority view which might now prevail and that which would have prevailed fifty years ago."[4]

While Mr. Conant's argument may seem intuitively plausible to most of us, it raises a host of problems of definition. "Accumulation" might mean, for instance, an increase in the number of items of scientific information in certain lines of endeavor, with no decrease or loss of such items in these or any other fields to offset it. In the actual history of science, however, we frequently find that a generation that has learned many new things in certain fields may at the same time have forgotten other scientific or technological data well known to their ancestors. Thus the engineers of the seventeenth century excelled by far the Romans in their mathematical skill, yet lacked their knowledge of how to build structures from concrete. Historians of medicine suggest that modern medical science may have paid for some of its advances in anatomy and physiology by partial and temporary loss of previously available knowledge in medical botany, drug and herb therapy, and what is now being called psychosomatics, that is, the study of the interplay of mental and emotional conditions with physical states of the body. Even on the level of practical design this writer has been shown by an anthropologist flint tools from the stone age which seemed to indicate a far more sophisticated understanding of the structure and function of the human hand than is applied in the design of the handles of most hand tools today.[5] The problem of evaluating "accumulation" in any kind of inventory to which certain items have been added while other items, though often smaller in number and significance, have been lost, has plagued students in many other fields, notably in discussions of such concepts as "biologic progress" and "economic growth."[6]

Answers to this problem of growth in a changing inventory or heritage have usually been given not so much in terms of sheer physical bulk as in terms of some over-all function performed with the help of the inventory or ensemble thus evaluated. In this way accumulation of growth has been measured in terms of increased probability of survival for individuals, groups, or species; or in terms of relative independence from wider ranges of different environments, by means of greater adaptability or greater power to transform them; or in terms of increased probability of reaching previously stated goals; or finally, in terms of increased learning capacity, openness to new ranges of information from the outside world, increased creativity, and greater ability to choose between widening ranges of possible goals. At this stage it should be sufficient to note the seriousness of the problems raised by all such concepts of "growth" or "accumulation," but we shall return to some of their aspects in a later section of this chapter.

Despite such difficulties of definition, it seems clear that Mr. Conant's distinction does touch upon some significant difference between the sciences and the humanities. Yet it is not certain that his distinction between "accumulative" and "non-accumulative" knowledge goes to the heart of the matter. It is true that Homer's *Iliad*, taken in and by itself, has never been surpassed, but neither have Euclid's Geometry and Newton's Physics been surpassed, if we take them as isolated intellectual achievements in and by themselves, or even as practical descriptions of the nature of things within the proper limits of their accuracy and verifiability today. It could also be claimed, on the other hand, that Shakespeare's treatment of the mother-son conflict in Hamlet has added something significant to the treatment of a similar theme in the Greek tragedy of Orestes.

Perhaps the essential difference for which we are groping may lie somewhere else. While it may be true that Shakespeare has added significantly to the treatment of a theme by Aeschylus, the fact of this addition or accumulation of contributions is of quite secondary importance from the point of view of art or the humanities, for these are concerned, as we have seen, precisely with judging and appreciating works of art or configurations of symbols as *integrated wholes*. On the other hand, the beauty or elegance or intellectual power of a great conceptual scheme, such as that of Euclid or Newton, is only of secondary importance from the point of view of science. What distinguishes the two fields of human activity is not the presence or absence of the accumulation of knowledge, or of a rich heritage of scientific, artistic, or philosophic patterns of symbols and backgrounds for the appreciation and recognition of new contributions. Rather it is the importance accorded to this accumulative element in the field of science, and denied to it within the field of humanistic effort.

Generally, therefore, we may conclude that in the sciences and in the humanities there occurs the constant accumulation of tools and data, of remembered patterns of information for prediction and discrimination, and that in the sciences and the humanities this accumulation of a growing heritage has to contend with the processes of forgetting, narrowing, trivialization, or other kinds of retrogression. Just as tools, data, and patterns of selection are cumulative everywhere, so problems, operations, and "working surfaces" in the new contacts with reality are everywhere new. The "wild surmise" on a peak in Darien was as new in the geographic exploration of Balboa as was that ascribed to Cortez in the sonnet of Keats. Cumulative and non-cumulative knowledge are thus two aspects of the same process of human thought, of the growth of human knowledge in operation, and the primary emphasis given to the cumulative aspect of this process in science, and the primary emphasis given to its non-cumulative or presentational aspects in the humanities, should not obscure this basic unity.

This emphasis on the interplay of cumulative and non-cumulative knowledge is operational rather than merely rhetorical. It would be difficult and perhaps impossible to appreciate many of the great contributions of literature, painting, or music, without at least some knowledge of the cumulative heritage of their particular fields. And it has been urged, as in David Hawkins' paper in this volume, that science cannot be adequately understood by an exclusive emphasis on its cumulative aspects, and that the qualitative and unique aspects of scientific situations are indispensable for its proper understanding.

The interplay of the repetitive and the unique, the cumulative and the non-cumulative, the discursive and the presentational aspects of knowledge is particularly characteristic of those disciplines of knowledge that seem to lie, as it were, in the middle between the repetitive and single-minded emphasis found in certain stages of mathematics and physics, on the one hand, and the presentational awareness of the creative artist in literature, music, or painting, on the other. In such fields as medicine, psychology, and the social sciences, an awareness of the simultaneous and combinatorial interplay of many factors is indispensable for the understanding of many of the potentially unique cases or situations with which these disciplines attempt to deal. History and political science lie perhaps somewhat further along the same continuum, since they are even more concerned with the interplay of multiple factors in at least partially unique situations, and draw upon accumulated experiences in their own fields in trying to estimate what aspects of such situations are likely to recur.

Problems of Scale in Time, Size, or Numbers

The problem of the varying emphasis on repetitiveness as against simultaneity as a criterion of relevance in various fields of knowledge, sketched in the preceding paragraph, may not be unrelated to the problem of scales of time and perhaps of size, as emphasized some years ago by Norbert Wiener.[7] According to this view, the physicist can deal with the behavior of molecules, atoms, and electrons because he limits his interest to their behavior in extremely large numbers. On this scale of vast numbers of particles and even vaster numbers of changes occurring among them, only the repetitive and statistical aspects of their behavior will stand out; and since the scientist has limited his interest to these particular aspects, he will calmly accept as a rule his almost complete inability to predict the behavior of an individual particle, or the outcome of a short sequence of changes among a small number of such particles.

In the social sciences, still according to Dr. Wiener, the opposite situation prevails. The doctor and the psychologist are critically interested in the behavior and fate of individuals; and the epidemiologist, the economist, and the student of politics are all interested in events which will occur within a very few years, or usually at most within the lifetime of one or two generations. The statistical regularities which thus predominate in the large-scale view of chemistry and physics, and even in much of biology, are quite unsatisfactory for answering the small-scale and short-term questions of social science, where human beings undertake to study processes whose detailed time-scale is perilously similar to that of their own lives.

On closer examination of the history of science, however, Dr. Wiener's argument might turn out to be less absolute than it seems at first sight. It is true that workers in the "exact sciences" have concentrated most of their efforts first on those processes which had a time-scale particularly remote from their own, such as astronomy and geology, and, secondly, on the behavior of large aggregates of physical particles or living cells. As these sciences progressed, however, they encountered an increasing number of problems involving time-scales closer to those of the human observer, as well as problems of relatively unique situations and decisions between different outcomes of different probabilities. Contemporary studies in the fields of meteorology, the management of flame in jet engines, the prediction of turbulent flow, the prediction of stress and breakdown in rigid structures, the analysis of the performance of instrument systems, and the interest of mathematicians in stochastic processes, games theory, and decision problems are all indications of this trend.

If this should be so, then it may turn out that the exact sciences have won some of their early successes by avoiding the kind of problems still too difficult for their yet undeveloped powers. They may, indeed, now be moving towards more intense preoccupation with problems involving the simultaneous interplay of a relatively small number of factors, and the consideration of scales of time and magnitude much closer to those of human life. If this trend should in fact materialize, we could hardly expect the qualitative and human aspects of science, and the entire atmosphere of scientific work, to remain wholly unaffected by it. Important aspects of science and scientific thinking, in this event, may become less dissimilar to some of the major concerns of our humanistic civilization.

At the same time some of the social sciences—as the medical sciences long before them—have been moving towards a greater concern for statistical probabilities both for larger and for smaller aggregates. As a result, sciences as different as meteorology and the study of the behavior of small groups in society are approaching a similar style of thinking. Even though it may be far too early to entertain any particular expectations of such a convergent development in the immediate future, it is perhaps possible to detect already indications of such a trend in some of the studies of certain qualitative aspects of science today; and a further study of these aspects of science may well prove rewarding for the student of civilization at the present time.[8]

II. SOME QUALITATIVE ASPECTS OF SCIENCE

The question has often been raised whether science can be discussed rationally by "outsiders," such as non-scientists in general, or even outsiders to each particular scientific field. What is more, scientific innovators have frequently complained that they have not been understood by the great majority of their fellow specialists in their own time. In the bitter words of Max Planck, "A new scientific truth does not triumph by convincing its opponents and making them see the light, but rather because its opponents eventually die, and a new generation grows up that is familiar with it."[9] There is something unique in every new creative moment, in every new situation of insight, and the essential loneliness of the pioneer is well known in the history of science. Since pioneering ideas in their pre-experimental stage may not be communicable even to scientists, a richer and more sophisticated appreciation of the processes of intellectual creation and creativity might be of some usefulness to scientists themselves. The autobiography of Walter B. Cannon, *The Way of an Investigator*, and the little book by Jacques Hadamard on the psychology of mathematical invention are well-known efforts by major scientists to explore these aspects of their work.[10]

If much of science appears to be more inhuman to the layman of today than it did during most of the eighteenth and nineteenth centuries, this may be owing in part to a qualitative difference in human ignorance. Laymen today, just as in past centuries, have at least some familiarity with middle-sized objects, such as billiard balls or houses (middle-sized, that is, in contrast with electrons or galaxies), or with simple human organizations, such as hierarchically ordered governments. From houses they can derive the notion of structure; from stones or billiard balls the notion of corpuscles or integers; from bureaucratic tyranny, the notion of hierarchic order. But there seem to be no obvious and familiar objects today from which the behavior of such entities as neutrons, mesons, electrons, or star-clouds can be made intuitively familiar to the layman, and the thinking of the scientist as well as of the creative artist must now suffer from the corresponding handicap.

Yet in their day the conceptions of a revolving earth, an invisible gas, an induced electric current, or a magnetic field, when first proposed, seemed as outlandish as some of the conceptions of modern physics seem to most of us at present. We are becoming increasingly familiar with patterns of change and of motion, such as the scanning techniques of television cameras, the coding and decoding processes of communication systems, the focusing and dissolution of photographic images, sequences of patterns in motion pictures, the changing patterns of traffic flow and of turbulence in liquids and gases, the processes of tuning, resonance, and modulation, the automatic matching and recognition of intricate patterns in Yale keys and punched cards, the convection currents in flames, and the intricate patterns traced by the movement of our own bodies in time and motion studies. There is reason to think that we have hardly begun to use the wealth of the new patterns with which we are becoming familiar. Certainly it seems too early to say that the work of the fundamental scientists has carried us into a world of processes too small or too large, too fast or too slow, ever again to be visualized and comprehended by our imagination.[11]

After we have decided how far and how long the discoveries of modern science may outgrow the resources for visualization and imagination available in our culture, we meet the question whether it is at all possible to convey the "feel" of science to non-scientists. Is there a fundamental difference between working in science and thinking about science, similar to the difference between the boxer and the boxing coach, the craftsman and the efficiency engineer, or the artist and the critic? Are such differences irreconcilable? Or are there ways in which human experiences can be and have been reunited?

The experience of human civilization over the centuries suggests that the differences between the actor and the observer, the doer and the critic, have been bridged time and again in the past. In the creative arts, as well as in the practical arts, in literature as well as in politics, common standards and conceptions have been re-established between the craftsman and the connoisseur, between the artist and his public, between the statesman and the citizen. In the history of science itself we find that the abstract numerical approach which was the mark of a few scientists and philosophers in the days of Pythagoras and Euclid became commonplace in the training of high-school students and engineers in modern times. Again, much of the world of Newtonian and Cartesian science became the common possession of broad strata of educated people in later generations. While, as we have seen, there seems to be clear evidence that no such real integration between the thoughts and experiences of working scientists and the mental universe of our general community has taken place today, there seems no reason to conclude that such a reintegration of the thinking and feeling of our civilization will not be achieved in the future. That at least partial reintegrations of this kind may even be probable is suggested even by our present rudimentary knowledge of the nature of human thought.

Rationality and Intuition in the Organization of Human Experience

Thought is based on memory, and on the use of symbols by means of which the recall of stored data of information is governed. Human memory is characteristically abstractive, dissociative, combinatorial, and pattern-seeking. We abstract items of information from the fullness of our sense-impressions, and store them as traces in our memory. We are able to dissociate different items of information from each other which were originally remembered as part and parcel of the same event, and we are capable of recombining these dissociated items into new configurations which were never found in experience before. Finally, we are able to abstract new simplified patterns from these new combinations—patterns so simplified as to obliterate the traces of their combinatorial origin—and to store these new patterns again in our memory, as well as to apply them to action.

What we can do in our memory as individuals, we can also do as members of an organized society, either among contemporaries or over longer periods of historical time. To the storage of memories in the mind of an individual there correspond in society the accumulated traditions of cultural life, civilization, science, or technology. Again we find at work the processes of abstraction, dissociation, recombination, the abstraction of simplified new patterns, and their verification and application to behavior. In the individual, as in the society, we find memory and tradition functioning as necessary conditions for initiative and creativity.

Men and groups of men can thus create new things and initiate new patterns of behavior. But can they retrace and retain what they have once initiated? In order to be made reliably repetitive, thought and knowledge must be retraceable at every major step. This is the function of *rationality*. Essentially, we may think of "formal rationality" as the marking of every significant step in a mental process, or of every significant stage along a trail of thought, by means of some secondary symbols or traces in memory, such that the sequence of mental operations can be retraced and repeated with substantially identical effects by the person who thought of it the first time, as well as by others. This we call the "cogency" of reason, or if carried to an extreme degree, the "rigor" of mathematical proof.

From marking and remembering the sequence of mental steps in particular operations, we may proceed to marking the pattern of the sequence of permissible mental steps for whole classes of mental operations. If we do so, we shall have abstracted rules for particular ways of reasoning or thinking. Such rules may be abstracted from the frequently observed sequences of events in nature, or from the repeated experience of particular patterns of argument within a social group.[12]

Obviously, the marking and making retraceable of a particular sequence of thought indicates nothing about the content of the thoughts themselves. In particular we are told nothing about their truth. The same applies to the laying down of rules for whole classes of mental processes, such as rules for "correct" logical thinking, or for arithmetical calculations, or for the correct playing of some particular game. All of these are formal or secondary conventions, which tell us nothing about the primary content of the mental processes which they are supposed to regulate. But if those rules have been so chosen as to correspond to observable regularities in nature, then they become powerful instruments for the deduction and calculation of real events.

To be sure, memories can be recombined and mental operations can be carried out without stopping to mark and remember each major step taken, or without taking note of the general rule by which their pattern could be described. Perhaps it is this that we have in mind when we speak of "lightning quick" or "intuitive" thinking. Thus the story is told of Pascal that he once gave immediately the four-digit result of a complex mathematical problem, saying that he would now have to work for an hour to find out how he had obtained it. Whatever credence we may give this anecdote, it is well known that mathematicians sometimes find new theorems first, and then have to work for weeks or months in order to establish proofs for them; and there are cases in the history of mathematics, such as that of certain theorems by Fermat, confirmed by all empirical tests made thus far, but for which no proof has been found to this date.

To the much greater speed and boldness of intuitive thought corresponds, of course, its lesser reliability. In the absence of repetitive checking of each major step, the opportunities for error are compounded with the length of the sequence of mental operations, or with the richness of the universe of memories that has been scanned.

In another sense, however, intuitive thinking can be extremely accurate. This is particularly the case in operations of recognition, that is essentially in the rapid scanning or presentational comparison of two different patterns for their over-all similarity or mutual fit. One of the two patterns to be compared is usually derived from some sense-perception of the contemporary environment; the other is usually stored and recalled from some type of memory—individual, social, or quasi-mechanized—as in the case of a photograph or a fingerprint. In most of these cases, however, an over-all comparison may lead to a relatively rapid "yes" or "no" judgment to the effect that the pattern to be tested does or does not sufficiently resemble the pattern used for testing, or meet the remembered standard. Such qualitative and sometimes even aesthetic judgments and acts of recognition occur in all fields of science, and they are even more familiar in the creative arts and the humanities in general.

The intuitive and the repetitive aspects of thinking may meet in the development of *style*. The repetitive aspects of style may be indicated, at least in principle, by something like a "recursion formula" in mathematics, that is to say, by a general rule stating at what points—or at what intervals or under what contingencies—a more or less closely characterized pattern is likely to recur. Thus the style and proportions of a Greek temple or of a Gothic cathedral may tell us just when and where a recurrence of columns or Gothic windows is to be expected, and what the general shape of these architectural elements is apt to be.[13] Similar though often far more subtle patterns of recurrence may be found in the style of paintings, poems, or forms of literature. Combined with these repetitive elements, there are found in style other elements of far-reaching recombination or even of relative randomness; and in the combination of the repetitive and expected, and the combinatorial and unexpected lie perhaps some of the most attractive possibilities of style. Style in this sense is found in scientific thinking and experimenting as well as in the most abstract operations of mathematics, and it finds its counterpart in the relative omnipresence of elements of style not only in the arts but in almost all elements of human culture.

It is against this very sketchy background of some of the common characteristics of human thinking and experience that the relative autonomy of science since the seventeenth century should be appreciated. It is with this context in mind that we may gain new insight from the study of its history, and that we may turn to the general descriptions of science given elsewhere in this book. The interplay of memory and recombination, of tradition and rationality, of calculation and intuition, of recognition and style, can all be found to a greater or less extent in the pre-Socratic science of the early Greeks, in the Aristotelian science of their later successors, in the scholastic science of the high Middle Ages, and perhaps even in the science of Descartes.[14]

Against the background of these common characteristics of human thought we may then recall the gradual emergence of the characteristic operations of modern science as we know it: the creation of new concepts by processes of dissociation, recombination, and recognition; the learning of new experimental techniques, often derived from the practical arts; the insistence on operational verification by repeated observations or experiments, and the steady improvement of methods and standards to that end; the ever increasing precision and multiplicity of measurement, with new concepts of science becoming embodied in new instruments of measurement, and with new levels of precision and significance of the measured data forcing the adoption of new scientific concepts; and finally the ever increasing insistence on verifiable predictions as the results of science. At one time, such prediction was thought of in terms of the absolute philosophic concepts of determinism and causality. Later, as these absolute concepts became obsolescent, prediction came to be thought of in terms of probability. Still more recently, prediction has been regarded simply as an operation: the abstraction of some patterns from some time series in the past, and their projection, singly or in combination, into the future.[15]

Every attempt at predicting the future in this manner implies an attempt to transcend the past by extrapolating its patterns beyond the limits within which they had appeared thus far. What applies here to the future in time, applies in a sense also to the very large or very distant in space, and even to the very small that may have remained thus far inaccessible to our observation. In trying to deal with any one of these, the scientist faces the task of transcending the limits of the known and heretofore observable by extending the patterns abstracted from experience, and transformed into patterns of questions, out into the unknown.[16] To the scientist, this attempt to transcend the ever receding horizon of his own knowledge is not a matter of choice but rather of unceasing necessity. The very foundations of his science teach him to assume the fundamental connection of events in the universe, and consequently to assume the merely provisional character of all isolated observations or experiments, and of all attempts at isolating, physically or intellectually, their objects. The provisional single-mindedness of the investigating scientist, and the creative artist's preoccupation with the unique, are both transcended by this overriding assumption of connection in science, and perhaps of philosophy or religion in the language of the humanities.

Reaching beyond the limits of present knowledge implies openness to new knowledge that may be gained from matching and remembering new patterns or items of information encountered in the unknown. It is by thus "taking information off the universe" that a scientist is able to add new knowledge to the established body of data in his discipline, and it is by fundamentally the same process that a calculating machine, such as an electronic chess player, could at least in principle learn to out-perform or out-play its designer.[17] In addition to learning from the universe by fetching new data out of the unknown, men further learn by recombining these new data with patterns, ideas, or data already in their possession in a universe of potential new recombinations, and by discovering and selecting within this potential universe those new combinations which, though improbable, are relevant to their needs. To some extent it may be said that the artist performs an important part of his work in a similar way. He senses, perceives, and selects new patterns from the world around him; he creates within his mind a vast universe of potential recombinations; and he selects from this universe, deliberately or by intuition, those patterns which appear to him as particularly relevant, even though they may appear highly improbable—that is, original—both to him and to others.

In addition to their improbability and relevance, most of the great symbols created by the artist and the scientist are "deep," in the sense that they are not easily exhausted. The inexhaustibility of possible significant meanings has been specifically acclaimed as a characteristic of great art,[18] and the possible multiplicity or fruitfulness of interpretations and applications of certain scientific concepts seems to be suggested by at least some writings in the history and philosophy of science.[19] In this sense of their potential inexhaustibility, or at least of the greatly extended validity and relevance of their possible meanings beyond those envisaged by their own originators,[20] some of the great works of art as well as of science grow beyond their own epoch into parts of a living civilization.

Some Special Aspects and Instances

What has been said thus far in general terms could be investigated more closely in the particular strategic cases or situations where science and civilization have encountered one another with significant results. Such situations may be found in some of the studies or documents relating to the psychological and biographical aspects of science, for example, those by Jacques Hadamard and Walter B. Cannon.[21] These documents contrast strongly with biographies of scientists written essentially from the intellectual outsider's point of view, such as perhaps the biography of Marie Curie by her non-scientist daughter Eve.[22] A remarkable instance of the use of an artist's image of a creative scientist in a scientific biography is found in Philipp Frank's biography of Albert Einstein, in which the author cites a long description by the poet Max Brod from a novel about the astronomer Kepler for which the young Einstein himself had served the poet as a model.[23]

Specific references to the aesthetic element in science may be found in some of the work of John Von Neumann,[24] and a serious attempt has been made by the mathematician George Birkhoff to give a mathematical formulation of certain aspects of elegance or beauty as a problem of minimization.[25] Instances of the importance of the intuitive element in science are familiar from the biographies and contributions of the mathematicians Pascal and Fermat, and perhaps from the initial belief of the elder Bolyai in the possibility of a non-Euclidean geometry, which was eventually vindicated by his son. A view of science as a succession of broad images of various aspects of nature, that is to say, of conceptual schemes in terms of which subsequent observations and experiments are conceived and executed—such as Torricelli's of the "Sea of Air," and the subsequent work on barometric pressure—has been made familiar by the writings of J. B. Conant, and has been successfully applied to the education of laymen in the fundamental principles of science.[26]

Literary appreciations of science, on the other hand, revealing both the power and the limitations of the humanist approach to scientific processes, may be found in the famous description of the Laputans and other misguided scientists in *Gulliver's Travels*, and, in a more optimistic vein, in many works by H. G. Wells. Perhaps a more penetrating humanistic view of science has been offered by Leo Tolstoi, and, following him, by Aldous Huxley, who suggest that science serves as a kind of stress test that brings out the hidden implications of the human or social institutions by increasing the power at their disposal. Science thus can make a good society better but it is certain to make bad or unjust social institutions worse.[27]

Some Problems of Science and Values

With Tolstoi and Aldous Huxley, and to a lesser extent even Swift and Wells, the focus of interest shifts from the aesthetic to the ethical aspects of science. Science itself depends for its life on the prior acceptance of certain fundamental values, such as the value of curiosity and learning, the value of truth, the value of sharing knowledge with others, the value of respect for facts, and the value of remembering the vastness of the universe in comparison with the finite knowledge of men at any particular moment. Historically, such values have been held by outstanding scientists. One thinks of P. W. Bridgman's well-known dictum that—"in the face of the fact, the scientist has a humility almost religious"; or of Newton's description of his own work as the play of a child with pebbles on the shores of the ocean of knowledge, or his reference to the sharing of knowledge with others by describing his own achievements as being due to his having stood "on the shoulders of giants."[28] Beyond such evidence, it could perhaps be shown that the cumulative work of science could not go on if any of the values just listed were rejected.

As science rests on certain values, so do almost all values depend on knowledge, and thus to some extent in turn on science, if they are to proceed from the realm of words to that of action. This implies a circular chain of causation or a feedback process, as do many processes of social and cultural development.[29] To act morally is in one sense the opposite of acting blindly. It is acting in the presumed knowledge of what in fact it is that we are doing. Almost every significant action of this kind implies serious assumptions in some field of science. To love one's neighbor requires at the very least that we find out where and who our neighbor is. If we are to respond to his needs we must first ascertain what his needs are and what action in fact is likely to be helpful to him. To feed the hungry requires first of all the ability to distinguish food from poison, as well as the ability to provide food or produce food when needed. The same principles apply, of course, to clothing the naked or healing the sick. Indeed, it can be said that perhaps no action can be evaluated as good or bad without some knowledge or surmise about its consequences. If we evaluate an action as good on the basis of our mere surmise of the good will of its doer, therefore, we may find ourselves forced to assume that such subjective good will—as in the Kantian Imperative—must include by implication also the will to gain and apply the best available knowledge of the probable consequences of the action chosen. The duty to have good intentions, in other words, is meaningless without the duty to try to know the facts and try to foresee correctly the consequences of one's deeds, and it is this latter duty which may distinguish in practice the responsible from the irresponsible statesman, or the well-intentioned doctor from the well-intentioned quack.[30]

Attempts at hermetic separations of science from values are thus bound to fail. Science without at least some values would come to a dead stop; ethics without at least some exact and verifiable knowledge would be condemned to impotence or become an engine of destruction. Much of the anxious discussion of international politics between statesmen and atomic scientists, or between the so-called schools of "idealism" and "realism" among political writers, hinges upon the discrepancy between the strength of the moral convictions involved and the poverty of reliable knowledge of the probable consequences of the proposed courses of action.[31]

The relationship of science and values thus implies a double question: the mutual interrelation of science and the general values of a civilization; and the relationship of a specific state of scientific knowledge to the pursuit of specific purposes or policies. The first of these problems, the general relationship of science and value, and thus to some extent of truth and goodness, leads us close to the heart of every civilization within which it is examined. If conceived as mutually incompatible, science and values may frustrate or destroy each other, dragging their civilization towards stagnation or decline. As a mutually productive and creative partnership, science and values may succeed in strengthening each other's powers in a self-enhancing pattern of growth, rendering their civilization increasingly open and able to learn from the hopes and dreams of the individuals within it, as well as from the universe around it.

This general vision of a mutually beneficial partnership becomes increasingly difficult to retain, however, as we proceed from the consideration of the growth of civilization on the grand scale to the effect of the timing of particular discoveries or innovations upon specific policies at specific times and places. Would it have been better for mankind if Einstein's principle of relativity, or Chadwick's discovery of the neutron, or Hahn's work on uranium fission had all come ten years later than they did, and no atom bomb had been available to drop on Hiroshima? Perhaps the most useful consideration in the face of questions such as these might be to realize the impossibility of foreseeing the ultimate consequences of even the smallest scientific or technological advance, as well as the inexhaustibility of most or all of the great contributions. Benjamin Franklin's answer to the question "What is the use of a scientific discovery?" consisted in asking the counter-question "What is the use of a baby?" Just as it seems impossible to foretell the eventual good a child may do, so it is impossible to foretell what evil he may do, and our whole attitude to children is in a sense based upon the bet that the good they do will far outweigh the evil. In civilized countries we have long ago abandoned the discussion which sometimes still echoes in mythology, whether a certain child should have been killed at birth in order to forestall the harm he did in adult life. Rather we have come to center our attention on providing a family and an environment for him in which love will outweigh hate, and in which his opportunity for free and friendly growth will be the best.

If there is merit in Benjamin Franklin's argument, we might similarly decide to bet on the potential goodness rather than on the potential evil of knowledge, and concentrate on providing a human and social environment for science in which its constructive possibilities are likely to be realized. It is possible, of course, to imagine extreme situations for some times and places in which the short-term potentialities for destruction might seem so great in the case of a particular invention or discovery, and the prevailing political régime might seem so unlikely to avoid its suicidal misuse, that a policy of temporarily restricting, delaying, or withholding such knowledge might appear as the least of several likely evils for the time being. Even granting all these assumptions, however, such a policy of fear of knowledge would have to be viewed as extremely transitory and exceptional in any modern technological civilization that is to continue to advance or indeed to survive. A civilization so prone to commit suicide that it could be saved only by concealing from it the means of its own destruction would not endure for long. Rather, for the long run and for most conditions that are likely to occur, we might do better to adopt the opposite assumption: that any modern civilization that is to endure will have to learn how to live with its new knowledge of its vast means of destruction.

III. SCIENCE AND CIVILIZATION AS MUTUAL RESOURCES AND CONSTRAINTS

Even this brief glance at ethical problems has raised the question of the whole relationship of science to the civilization of which it forms a part. We may thus think of the humanistic sectors of civilization as a resource for the development of science, as well as a possible constraint upon it. Conversely, we may think of science as a resource for general culture and civilization, as well as a constraint upon certain humanistic or cultural developments. All of these relationships may be found intermingled in particular cases in the development of science or civilization, and something about each of these inferences may then possibly be revealed by investigation of them. Finally, an understanding of the interplay of science and civilization in their double effects as resources and constraints for each other may help us to gain some clues for the understanding of the process of growth in nations, cultures, economic systems, and whole civilizations.

None of these topics can be presented here with anything like adequate treatment. All that can be attempted in the rest of this chapter is to indicate something of the significance of these relationships; to refer to at least a little of the abundant material available for research on these matters; and to suggest the potential significance and fruitfulness which the study of the humanistic aspects of science, and hence of the interplay of science and civilization, might gradually gain in the future.

In arranging our data for this discussion, we must agree on some scheme for their ordering, however rudimentary. Accordingly, we shall first discuss some effects of humanistic civilization upon science, and later some effects of science upon the humanities. Among these effects, we shall discuss in each case resources before constraints. Among the humanities, we should begin with those which seem closest, at least in some aspects, to the step-by-step discursive approach of the traditional sciences, and we shall then proceed to those of the humanities which are increasingly presentational in their emphasis. All such arrangements, of course, will be highly imperfect; they correspond to the attempt at representing a complex landscape by a sequence of symbols strung out along a single line. Yet some such sequence must be chosen in order to fit the requirements of the printed page. Without taking our arrangements too seriously, therefore, we shall discuss among the humanistic areas of interest, first of all philosophy and religion, then prose literature and poetry, and finally the arts and music. Among the social sciences and arts, we shall similarly proceed from economics and business activities to sociology and what the anthropologists call culture, and we shall end up with the field of historical and political studies, since these deal at least in part with the making of enforceable social decisions in the light of all these other aspects of human activity.

The richness of humanistic resources for science is, of course, much greater than any system of mental pigeonholes we may contrive in which to file our data. Within any pigeonhole, again, we shall do little more than indicate a very few examples of the kind of humanistic resources that could be traced in the past and present life of science, and which might offer rewarding opportunities for further research.

Humanistic Resources for Science

The extent of the resources of philosophy and religion for the growth of science is perhaps indicated by the fact that these three fields of activity are often inextricably intermingled in the early stages of civilization. Even after philosophy had become separated from religion, and science from philosophy, such fundamentally philosophical notions as coherence, consistency, truth, as well as quality, inclusiveness, understanding, and others, received a great deal of their sharp formulation from philosophers, although these concepts often had been taken from various fields of everyday life and later applied in the working and thinking of scientists.

The traditions and injunctions of religion have long served as a storehouse and resource for much of the psychological and medical knowledge of mankind. Religious cleanliness taboos and dietary laws have represented much of man's early knowledge of public health, epidemiology, nutrition, and perhaps of psychosomatic medicine. The approach of Hippocratic medicine with its assumption that every illness represented a condition which in principle could be traced to sharply identifiable natural causes, was perhaps most successful in those types of illness where a few readily apparent physical conditions turned out, on further examination, to be linked with a high degree of probability, as "symptoms," to an easily recognizable disease pattern. Where, on the other hand, the initial conditions or causes were difficult to define or recognize with the means of a primitive technology, or where they were connected with their overt consequences only with much less degree of probability (for instance, by no means all eaters of pork could be expected to develop trichinosis), "irrational" or religious formulation of rudimentary medical knowledge has been much more persistent.

In another context, religion has been a major storehouse of resources for man's knowledge of psychology and of its application to the lives of groups and individuals. Thus, classic Greek science and philosophy treated children essentially as objects to be moulded into desirable types of grown-ups by the philosophers or by the state, or merely as resources for manpower, civic morality, or excellent or virtuous performance in terms of previously laid down standards; Christianity produced a radical change in this way of thinking. For Christianity, children became possible sources of initiative, beings with a far higher degree of openness and learning capacity than could be found ordinarily in grown-ups, and thus potentially greater recipients and greater creators of values. Where the Greeks had endeavored to make children like grown-ups, or more specifically like the most desirable type of grown-ups they could think of, Christianity asked grown-ups to become "as little children." While much of this emphasis was overlaid later by other concerns in the development of the medieval church, it was at least likely to be rediscovered with the new interest in the fundamentals of early Christianity in connection with the eighteenth-century concern with "natural religion." Thus, if Rousseau has been hailed by some as one of the pioneers of modern psychology and modern education, it is perhaps no accident that he was also the author of the liberal religious views of the "Savoyard Vicar"; nor is it perhaps an accident that the age that produced "Jefferson's Bible" and the views of the Deists also produced the educational ideas of Froebel and of Pestalozzi.

Prose literature and poetry represent another major resource of scientific thinking, not only in the fields of psychology and education, as in the case of the many psychological and educational novels from Goethe to Marcel Proust and Thomas Mann, or the use of the theme and symbol of Oedipus in drama and poetry, long before its use by Sigmund Freud and the psychologists, but also in such compelling visions as the Icarus myth and other poetic images of human flight, which foreshadowed actual scientific and technological developments by almost three thousand years. In passing, we might ask whether possibly modern students of logic and semantics may not have learned something from the rich and subtle tradition of poetry in conveying precise shadings of concepts, meanings, and associations. The cultural tradition which permitted the development of the poetry of Rainer Maria Rilke and Karl Kraus may have been a resource for the precise thinking of Rudolph Carnap and Ludwig Wittgenstein, and a similar connection might some day be traced from the intellectual attitude of Baudelaire and symbolists whom he inspired to the stress on the symbolic character of science in the thought of a mathematician like Henri Poincaré. We may further ask whether poetry has not been a major area within which both semantics and linguistics, among other sciences of man, have developed.

The fine arts contributed to scientific thinking such concepts as elegance, beauty, simplicity, perspective, symmetry, shading, hue, and color, and many more. Music in turn has contributed such concepts as harmony, rhythm, resonance, and derived from these perhaps such notions as "off-beat" and "out-of-phase," as well as perhaps the concept of modulation. From this point of view, it is intriguing to trace how a simple pattern is abstracted from nature in everyday life, such as a wave pattern from the ripples of the surface of a pond; how it is then modified with the help of such concepts as rhythm and modulation from music and the fine arts; and how it is finally developed into some predictive and verifiable conceptual scheme in some field of science.[32]

Social resources for science have perhaps received more attention than the specifically humanistic resources which we have just discussed, and in particular the importance of economic resources for the development of science has been the object of some study. We know something of the effects of economic demand, and perhaps of the more important effects of economic resources, on the probability and timing of certain inventions and discoveries. There exists a considerable literature on the importance of adequate financing of research and development, and on the importance of adequate resources for the process of social innovation by which the results of such research are eventually adopted into social and technological practice.

Perhaps less attention has been given to the importance of specific business practices as a potential resource for scientific development. The growing practice of accurate bookkeeping and quantitative calculation in European business life from the thirteenth century onward may well have contributed to the acceptance of a more efficient number system, to the gradual rise of a social and psychological climate of what Max Weber in a different context has called "calculating rationality," and to the creation of a social pool of calculating skills and skilled individuals, among whom prospective scientists could be recruited.[33] In a similar way, perhaps, a possible relation between inventory and taxonomy could be investigated; to what extent did merchants make and keep orderly inventories of their goods before scientists began to compile specific inventories of human knowledge in particular fields? And to what extent did both merchants and scientists learn the arts of inventory and taxonomy from earlier social organizations, such as monasteries or governments?

Other very different resources for science have come more directly from sociological patterns, such as the existence of particular classes of bards, minstrels, or scholars, and the respect paid to them by the rest of the community. The high social status accorded to the Talmudic scholar in Judaism, or to the Brahmin in Hinduism, or to the scholar in traditional Chinese culture, or to the university *Professor Ordinarius* in nineteenth-century Germany, may not be unrelated to the theoretical contributions made by members of these cultures to the science and philosophy of the world.

In some of these respects, the sociological pattern of particular groups may be replaced or reinforced by cultural patterns of tradition or preference which may be more widely diffused throughout the society. Such cultural values conducive to the growth of science may well have been values placed on scholarship, on literacy, on rationality or retraceability of arguments, and in particular perhaps on the cogency of arguments among equals, as against a mere acceptance of propositions from status or authority. It has been argued, for instance, that the attempt of the medieval cathedral builders and theologians to make a large part of Christian doctrine capable of visual demonstration to the believer has in turn contributed to the attempt to make as much as possible of human knowledge impersonally demonstrable in all fields. This, in turn, is a recognizable characteristic of much of modern science.[34]

Some of these social and cultural resources for science are reinforced by the resources of government and public administration. Thus both the interest in and the art of gathering, verifying, and using quantitative data on a large scale seem to have been developed for governmental purposes long before they were discovered to form a powerful tool of the scientist. The administrative records of births and deaths turned out to be the essential foundation for the compilation of rate of mortality tables, insurance statistics, and other fundamental steps towards the development of statistical probability and prediction.

Humanistic Constraints on Science

Philosophic concepts have often been constraints on science. A radical insistence on philosophic idealism, and a corresponding devaluation of the observable world as essentially a mere collection of illusions and fleeting and fundamentally unimportant phenomena, is not compatible with the expenditure of a great deal of attention and energy on the making of such careful sense-observation as scientific work demands. The trend away from observations in classical science after the age of Plato has often been commented on, although Plato's influence does not seem to have prevented the making of some excellent observations by his disciple Aristotle. If Plato's philosophy did have a constraining effect on the development of empirical science, though not necessarily on that of mathematics, the evidence does not seem quite clear as to whether the constraint was due to Plato's distrust of fallible, sensory experience, or to his very considerable confidence in the possibility of obtaining true knowledge more directly and with greater certainty through an act of philosophic vision.[35]

In its common and somewhat ambiguous usage, the term "philosophic idealism" implies perhaps two different operations: the criticism and distrust of all sensory evidence, and the positive belief in a superior method of obtaining knowledge through more or less mystical intuitive insight. Even if we have learned to distrust observable phenomena, we might still investigate them with great care as long as we believe that they are substantially the only access to truth we have; only when we become persuaded that there is a better and more direct way to relevant knowledge which does not involve observing the world around us are we ready to turn away from observation and experiment. Kant in the eighteenth century retained a good deal of the Platonic distrust of observable phenomena, but since he no longer shared Plato's confidence in attaining direct visions of the world of immutable ideas, his philosophy still left his followers with the sceptical but cheerful study of phenomena as the most promising road to knowledge, and in nineteenth-century Germany this philosophy appears to have been held by a good many scientists and to have proved entirely compatible with rapid and substantial scientific growth—a point which Marxist critics of "philosophical idealists" seem often to have overlooked.

Other philosophic constraints have been more specific, such as the constraints imposed on scientific thinking by particular philosophic assumptions of cosmic harmony or even of the perfection of the circle. The latter point is perhaps illustrated by the well-known statement of Copernicus that "the mind shrinks with horror" from the thought that the orbits of the planets could be anything but circular—a constraint that was only overcome about a century later by the calculation of elliptical planetary orbits by Johannes Kepler.[36]

The effect of some of the religious or moral constraints on science may have been even broader, such as, perhaps, the effect of the traditional Jewish and Moslem prohibitions of the making of images, or the religious hostility to the practice of dissection found in certain stages of Christian as well as pagan religious history, or the continuing opposition to vivisection in our own day. Wherever the authority of religion has been put behind the blind acceptance of unverified assumptions or the exhortation of authority above any process of reasoning or impersonal verification, we are apt to find at least some instances of the familiar "warfare between science and theology," although in the long run most of the world's great religions appear to have succeeded in adjusting to the successive interests and discoveries of science without any fatal loss to their own central religious traditions.

Literature may sometimes exercise a constraining effect on the prestige of certain scientific ideas and thus on the likelihood of their attracting other scientists to test or elaborate them. Thus Voltaire's attack on the prestige of Leibniz and Jonathan Swift's ridiculing of Leibniz' *Ars Combinatoria* in his caricature of the combinatorial researches of the scientists at Laputa may have contributed to the subsequent long neglect of many of Leibniz' ideas. Plutarch's comment in his "Discourses at Table" on the unworthiness of arithmetic as a topic of study for gentlemen may similarly have contributed to the subsequent direction of public and educational attention away from that field. In poetry, too, the habit of appealing to the authority of Homer, or the bitter and effective ridiculing of science by Aristophanes in some of his plays may have had a constraining effect on the subsequent development of science. Beyond its content, moreover, a work of poetry may involve a conflict with science because of its intrinsic beauty or formal appeal; thus the spurious Rowley poems actually written by Thomas Chatterton opened a conflict between those impressed by their beauty and those intent on verifying their authenticity or historical truth. The conflict between aesthetic and scientific methods of verification, and between the beautiful but possibly untrue myths and the true but possibly unbeautiful findings of research, has been a recurrent theme in historiography as well as in political and social science.

In this manner, also, the fine arts—or rather the standards implicit in the appreciation of fine arts—may function as constraints on science. Thus Lucian in his dream of Techne and Paideia comments on the ugliness of the sculptor who fashions beautiful marble sculptures with his hands; sculpture and all technology and work with one's hands, Lucian concludes, are apt to have inaesthetic effects on the body and mind of the person who does the work, and should therefore be avoided. Similarly, Swift's ridicule of the scientist who tries to study problems of food recovery from human organic waste gains much of its point from the reader's expected aesthetic revulsion from such an object of study, apart from the merits or demerits of the particular experiment satirized. In any case, the subsequent development of science showed that the study of organic fertilizer has been actually conducive to that growing of two blades of grass in the place of one of which Swift otherwise approved. Perhaps even music may have contributed to the number of temporary aesthetic constraints by its convention of harmony and definitions of dissonances which for a long time excluded atonality, and may have precluded in the West an earlier analysis of the scales and harmonies of the Eastern, for example, Indonesian, music, perhaps with a resulting lag in our understanding of these cultures.

In contrast to such humanistic constraints, many of the social constraints on the development of science are much better known. A great deal has been written about possible constraints on inventions through such economic practices as monopolies or restrictive patent policies, and the possibility of similar constraints even on geographic discoveries has long been discussed in connection with such topics as the question of any possible Portuguese mariners' maps of the Western world before Columbus. A number of potential economic constraints on organized research, application and development, and social innovation have been studied. Company policies may exclude or restrain research in certain areas. Ample land resources may have held back in some countries research on soil conservation, and may even have contributed for a time to the loss of existing knowledge and practices in this field. Abundance of timber similarly may have deterred man from the early development of adequate studies of ecology and techniques of forestry. Slavery has been widely discussed as an effective constraint on the development and introduction of labor-saving machinery, and notice has been taken of the lesser but similarly constraining effects, under certain circumstances, of an ample supply of cheap labor. Other economic constraints in the form of lack of capital may retard the spread of innovations, or may direct energies towards more immediately profitable speculative ventures which turn out to be scientifically and technologically sterile.[37]

Certain sociological patterns, such as particularly sharp barriers of class and caste, may sometimes function as even more effective constraints on scientific development. Thus the segregation and the low social standards of those who do physical work, and are thus in a position to have many of the empirical experiences on which so much of science feeds, have often been identified as at least a contributing cause to relative scientific stagnation. The classic Greek and Roman contempt for labor as the mark of slavery, or the more recent nineteenth-century attitude towards certain kinds of labor as menial, the segregation of the "untouchables" in India—all are likely to have slowed down empirical scientific development in the areas where these social practices existed. Cultural patterns extolling authority and deprecating contradiction or debate may have had similar effects. The well-known motto on the arms of the English Royal Society, *nullius in verba*—on the words of no man—illustrates the change in cultural climate which the unhindered growth of science required.

The restrictions imposed from time to time on science by governments are perhaps to a considerable extent the expression of the constraints implicit in the cultural and social patterns upon which these governments rest. Within each particular society, economy, and culture, however, each government usually also has a certain sphere of political discretion, and governments have sometimes used this discretion to make the constraints on science still narrower than they otherwise would have been. The political repression invoked against such scientists and humanists as Socrates, Galileo, Giordano Bruno, and others is too well known to need reiteration.

The more thoroughly a government has become smitten with the "disease of orthodoxy," the more harmful, as a rule, have been the effects upon scientific growth.

This does not mean, however, that science can grow only under democratic governments. It grew vigorously in the thirteenth and fourteenth centuries despite the inquisition in Germany, Italy, and France; and it grew even more vigorously at the courts of the Renaissance tyrants and in the absolutist states of the seventeenth and eighteenth centuries. Even though the Nazi dictatorship in Germany destroyed much in German science, it did not prevent the making of major contributions by Otto Hahn and others in fundamental science, as well as in applications of science to technology. Again, despite the serious effects of the Russian dictatorship and its heavy-handed interference in such fields as physics and biology, the peoples of the free countries have been frequently warned by their governments not to underestimate the actual extent and the continuing possibilities of scientific and technological growth behind the "Iron Curtain." The available data on the development of atomic energy, jet propulsion, and scientific and engineering manpower in these countries point in the same direction.

An interesting observation has been made in this connection by Philipp Frank. He finds that even dictatorial governments, as a rule, have not been eager to seek conflict with the unanimously or almost unanimously accepted findings of any particular science or profession. It is rather when scientists in a particular field have become divided among themselves, because there does not exist at the moment any clear-cut theoretical concept or conclusive experimental evidence to decide some particular question of major importance, that governments are tempted to rush in to fill the apparent vacuum in scientific knowledge by means of a political decision. Thus the attempt of the authorities in church and state to decide the questions of planetary systems occurred during the period between the work of Copernicus and Galileo, that is, during the period in which there was not enough definite experimental observational evidence for a decision on the matter. Similarly, according to Professor Frank, the attempt of the Russian dictatorship to decide questions of biology and genetics by a reference to the "party line" may have been occasioned by the inadequacy of present understanding of many aspects of the interplay of heredity and environment, and by the current lack in all countries of adequate scientific concepts and experimental data to decide these questions. In most cases, political interference is, of course, unlikely to accelerate the finding of such scientific concepts and techniques; as and when these are found, however, and the scientists have agreed on the facts of the matter, governments usually tend to accept or at least to tolerate their decision.[38] The study of these and other humanistic or broadly cultural and social influences on the outcome of particular scientific conflicts or decisions offers a fascinating field for research.

Science as a Resource of Humanistic Civilization

Just as humanistic concepts and practices have been a resource for scientific thinking, so scientific data and methods have offered ever new resources to humanistic culture. Religion and philosophy have both drawn inspiration from the new orderly universe of astronomy, so vastly greater and so vastly more predictable and lawful than the complicated and finite universe of mythology and of the Middle Ages. To medieval man God's rule might be just, but its workings could not be predicted. It was a modern question to ask whether God, like a constitutional monarch, could be relied on to keep his laws and not to suspend their operation by frequent miracles. "The eternal silence of these infinite spaces" still frightened Pascal. But after Newton had formulated a law by which it was governed, the emotional impact of this new experience may be felt in the well-known lines of Addison's hymn "The spacious firmament on high . . ." and in Kant's famous reference to "The starry sky above and the moral law within." The emotional impact of the discovery of the two new worlds of the infinite and of the infinitesimal has contributed to religious experience; and the experience of the steadily receding horizon of human knowledge has led a modern philosopher like Karl Jaspers to an essentially religious conception of the "encompassing reality" behind it.[39]

Even some of the most specific methods of science have been used as resources for religious thought. Thus Leibniz' religious conception of the "best of all possible worlds" is based in part upon the intellectual approach of the calculus, and upon the concepts of the maximum, minimum, and optimum, which it entails. Attempts by scientists to find in their science a confirmation for religion have continued from Isaac Newton to Lecomte du Noüy; and on the side of religion similar attempts have continued, from the use of Aristotelian logic by the scholastics to the "natural religion," Unitarianism, and Christian Science of the eighteenth and nineteenth centuries. In a broader sense, the cosmologists' views of the development of the universe, and the biologists' views of the evolution of life, have had an unceasing fascination for the literate mind, and have influenced in subtle ways the religious thinking of our time.

In specifically philosophic thought, the influence of scientific data and methods has perhaps been even stronger. From the seventeenth to the nineteenth century much of philosophic discourse developed in the "scientific" manner, that is in the style of a causality or necessity that could have been borrowed from physics, and with an insistence on demonstration that often was borrowed from Euclidean geometry. Spinoza specifically undertook to write an "Ethics in the Geometric Manner"; and in our own time the interplay of mathematical, logical, and philosophic thinking in the work of such philosophers as Bertrand Russell and Alfred North Whitehead has been conspicuous. The biological concepts of natural selection and of evolution have appealed to such different philosophers as Friedrich Nietzsche, Henri Bergson, and John Dewey; and a good many philosophic contributions in our time have been made by men who have also made notable advances in particular scientific fields, such as William James, Bertrand Russell, P. W. Bridgman, Charles Morris, Philipp Frank, and Norbert Wiener.

In the field of literature, physiological and medical data have been used to outstanding effect in such novels as Thomas Mann's *The Magic Mountain* and Sinclair Lewis's *Arrowsmith.* Serious philosophical novels have been written which draw many of their outstanding symbols from science and "science fiction" such as C. S. Lewis's *Out of the Silent Planet* and its sequels, Franz Werfel's *Star of the Unborn*, George R. Stewart's *Earth Abides*, and Albert Camus' *The Plague.* A number of novels have come out of the experience of flying; and science and technology have inspired some of the outstanding poetry of our age, such as W. B. Yeats' "An Irish Airman Foresees His Death." A biological symbol has been used to great effect in the poem by D. H. Lawrence "The Elephant Is Slow To Mate." A striking combination of the classic myth of Perseus and Medusa and the symbol of optical perspective in a view-finder is given in the poem "The Image" by Cecil Day Lewis.[40] The essay in this volume by F. E. L. Priestley cites an abundance of such examples.

Similarly, the methods of psycho-analysis and psychology have become a resource for the "stream of consciousness" technique of many modern novelists; and the ancient scientific techniques of collecting and taking inventory were used with hilarious results in the prose of François Rabelais and the ballads of François Villon, as in the "Ballad of Slanderous Tongues" in the *Great Testament,* where Villon proposes a long and learned list of unappetizing chemicals, pharmaceuticals, and other substances in which the tongues of informers are to be immersed.[41]

Scientific data and methods have long been a major resource for the fine arts and architecture. Much of the beauty of Gothic cathedrals came from such technological innovations as the use of flying buttresses to carry a share of the weight of the roof, which allowed an extensive use of ornamental glass in the walls. Le Corbusier suggests, further, that the beauty of Notre Dame was achieved by use of a mathematical concept of numerical proportions.[42] Present-day techniques of analysis and abstraction in modern painting, the use of juxtaposition, cutting, and close-ups in motion pictures, and the enlargement of detail in modern art books are other examples of the artistic application of scientific techniques. The use of parabolic and other curves in the construction of modern bridges by Robert Maillart, and the wide use of streamlining, flow lines, and similar curves have added new resources to the vocabulary of our design.

The development of science has opened up a new world of unusual experience to us from the world of the very small, revealed by the microscope and electron microscope, to the world of the previously unimaginably vast, as in the pictures of star clusters and galaxies now seen with the aid of the telescopic camera. An important study of this entire subject, entitled *The New Landscape,* by G. Kepes, is explicitly devoted to discussing some aspects of the impact of this new visual experience upon our world of thought and feelings.[43]

In the world of music, science and technology have long played a major role. The medieval organ was a technological as well as an artistic contribution, and so was the whole family of instruments that made the advanced music of the seventeenth and eighteenth centuries possible. Kenneth Conant has pointed out that elementary principles of acoustics were applied in many churches of the Baroque period, which were often explicitly designed to facilitate effective preaching. In more recent times, a machine technology has contributed to music new experiences of rhythm. The measurement of more accurate intervals, both in terms of rhythm and in terms of the intervals between tones, has facilitated the spread of syncopation and the experiments with the twelve-tone scale. New techniques in the use of microphones, of recording, and of sound cutting and combining, have made it possible for soft-voiced singers to be heard by millions, and have placed an almost unexplored variety of techniques at the disposal of the composer.

Scientific methods and data have offered perhaps even more conspicuous new resources to social practices and institutions. If the economic theory of the age of Adam Smith and his school was heavily indebted to the equilibrium concept of seventeenth-century physics, later economic and social theories borrowed heavily from late nineteenth-century biology in the age of "Social Darwinism."[44] Mathematical methods, and in particular the mathematics of probability, became major resources for business institutions in the fields of finance and insurance. Social institutions and cultural patterns have been affected on an even wider scale. Geometry was applied to the design of palaces, gardens, and city planning, from the gardens of Versailles to the streets of Washington, and even to Jeremy Bentham's "Panopticon" and the model penitentiaries which followed it. Systems of automatic regulation were applied by James Watt to the governor of the steam engine and, less successfully, by British statesmen to the sliding-scale tariff on the imports of grain. Policemen found first in the accurate measurement of the Bertillon System, and later in fingerprints and automatic searching devices for their filed records, new resources for the apprehension of culprits. Electricity supplied power for the use of floodlights in the theater and motion picture, as well as for political parades and the illumination of public monuments, and for the use of the electric chair. Efficiency engineering and assembly line methods have spread far and wide through modern culture, while the modern blood transfusion has created new opportunities for blood donors to express in a practical as well as a symbolic way the fundamental sympathy of man for man.

The science of bacteriology and the practice of asepsis have brought a new cultural standard of cleanliness in matters of housing, clothing, and bodily contact, while vitamin therapy and the science of nutrition have done much to change our attitude towards food.

Other resources of science have offered mankind the technical means to face the problems of a Malthusian increase in population by means of planned parenthood. Whatever the number of our children, science has been perhaps a major cause of our changed attitude towards them, and of the changed relationships between the child and its family, and between the child and his peers—a change that has been studied in some of its aspects with extremely suggestive results by David Riesman.[45]

Other effects of scientific and technological changes as resources for new culture patterns have been even more widespread. Glass and the mirror, and the beginning of elementary optics in the high Middle Ages, produced not merely the scientific and philosophic theories of Robert Grosseteste who considered the universe as essentially composed of light, but also with a flood of books called "Mirror" of one thing or another which covered Europe and added the word "speculation" to several Western languages. More recently, successful methods of animal breeding and tentative generalizations of Darwinian biology have been misapplied to human families or races, and have thus provided an opening for an ominous biological or pseudo-biological strain in Western politics, as well as for a genuine long-range concern for eugenic problems.

Governments as well as private psychologists and sociologists have found many uses for statistical methods, for sampling and opinion poll techniques, for projective tests, and for sound recorders in interviews and counselling. Since governments are in many ways vast organizations for human communication, it may be expected that application of communication theory to political problems, in which some beginnings have already been made, will continue to increase in the future.

Science as a Constraint on Humanistic Civilization

Science, at least during the last two centuries, has been in some ways an effective constraint on the development of religion. The scientific insistence on the relevance of only that knowledge that can be communicated, shared, and publicly verified has tended to deny or to play down the importance of all that has been called "private knowledge," "religious experience," or "revelation." Even when a scientist and philosopher like William James admitted "the varieties of religious experience" into the realm of meaningful discourse, these experiences no longer furnished themselves the standard for such criticism, but rather became their object.

Earlier, during the seventeenth and eighteenth centuries, such men as John Locke, Isaac Newton, Benjamin Franklin, and Thomas Jefferson had all found it difficult or impossible to accept the religious concept of the Trinity, which seemed to contradict sharply the principles of scientific economy and of the unity of physical bodies and organisms, none of which could be imagined to be in two places at once, or to have at one and the same time both a separate and a conjoint existence. It should be added, regardless of the merits of the theological arguments on either side, that modern science has become increasingly familiar with patterns of reality to which the fundamental limitations of classic particles or classic organisms do not apply, and that the intellectual conflict as such may be losing some of its tension.

Many of the scientific constraints that operated on religion also applied to philosophy. Here too all knowledge that was not exact, rigorously and publicly demonstrable, and analytical rather than emotional, remains suspect by scientific standards; and the philosophy of two centuries, from Descartes to Hegel, shows perhaps some of the traces of this constraining influence. Emotional approaches to philosophic problems, such as that of Soeren Kierkegaard, or of some of the modern existentialists, with their stress on the feeling of anxiety, became suspect as "mere literature" to many philosophers who modelled their own work on the standards of science.[46] This attitude perhaps was foreshadowed in some of its aspects by Plato's often expressed hostility to poets, and by his inscription "Let none ignorant of geometry enter" above the door of his Academy.

The constraints imposed on literature by the influence of science change with changing trends in science itself. Seventeenth-century science, still largely modelled on Euclidean geometry and Archimedean mechanics, sought for immutable regularities and classic patterns; astronomy and physics gave the clue to the sequence of events in all realms of thought, and the mind of man was devised, as it was thought, for the unravelling of relationships that would provide ultimately a complete explanation of every occurrence. Thus writers generally sought more and more exact rules of style and generic form; logic and grammar would in the long run answer all problems, and the diversities of human behavior would be comprehensively explained by the clear light of reason. The classic rules of dramatic construction were resurrected from Aristotle, and accepted in the age of Corneille; the critic's role was that of corrector of errors, for he concerned himself with the discovery of faults, of irregularities which marred the perfect symmetry of the whole.

With the passage of time, science and literature both left the narrow realms of the classical heritage, and ventured on parallel paths into new areas of thought and feeling. The complexities of biology, with its numerous related fields, from the "vegetable statics" of Stephen Hales to the comparative anatomy of Vicq d'Azyr, produced countless difficulties for a strictly mathematical view of phenomena; while the rise of the documentary forms of literature, the novels of Daniel Defoe, the *drame bourgeois* whose theorist was Denis Diderot, the rustic poetry of James Thomson, all fused in the cult of ideal nature fostered across Europe by Jean-Jacques Rousseau, reflect the breadth of the new outlook and the depth of its influence.

Freedom was gained in certain respects, but the poetic stature of the older forms was not easily rivalled in the new age. The cult of objective facts brought new constraints with it; religious themes were no longer so universal in their appeal, and began to find new values in the picturesque and the aesthetic, while the supernatural in general, once a commonplace in all forms of literature, from folk-tale to epic, came to be handled with so much self-conscious awkwardness that it lost not only its magic but also its serious intellectual content. Standards of verisimilitude rose remarkably, sometimes even to the point of destroying all other critical viewpoints, as the new outlook gained strength and consciousness of its objectives and techniques.

The vigor of realistic criticism largely destroyed the sense of classic form, of beauty for beauty's sake, and much modern writing is limited to a casual style, sometimes described as a technique of "multiple hedging." The sequence is not without its interest: the barbed and winged *Maxims* of La Rochefoucauld and the poetic eloquence of Racine gave way gradually to a stenographic record of the way in which men spoke, and a photographic record of their appearance and environment; and the diffuseness which tried to put the entire contents of a room into the pages of a novel, or the society of an epoch into a cycle of romances, gave way in turn to still another scientific ideal, that of rendering with exactness and economy the precise sequence of significant events. Balzac led to Flaubert, but after Flaubert there have been losses, in philosophic richness, in breadth and diversity of sympathy; the cult of impersonal treatment of highly individualized personalities has led to a literature in which the document, the case history, is supreme, in which the richness of the world is given us in a stream of consciousness.

In four centuries, the historic scope of Rabelais, Montaigne, or Shakespeare has been supplanted by the minutiae of James Joyce and Proust, and the interested observer cannot but ask if the atomizing influences of scientific development have not helped to produce the change in outlook and technique. A compensating development is suggested by recent trends in reading habits; the current vogue for literature of fact, describing political and social trends, ways of life in remote and unfamiliar parts of the world, adventures on balsa rafts or Himalayan peaks, indicates a dissatisfaction with literature as it has developed in the traditional forms. For other readers, almost the only form in which the novel still stirs the deepest fibers is that of science fiction; from Jules Verne and H. G. Wells through C. S. Lewis and A. C. Clarke, Ray Bradbury and Isaac Asimov, a sense of man's place in the totality of the cosmos is best conveyed by long-range travellers through space and time, meeting intelligences and forces that everyday experience can in no way duplicate. Here as elsewhere the influence of science is two-edged; as it opens the realms of the imaginable, it destroys the forms in which man has learned to create and recognize the beautiful and the good, and imposes a long and arduous effort to regain the serenity in which the artist may create once more in depth.

Apart from these developments, however, much of the effect of modern technology has been in the direction of constraints. S. I. Hayakawa has suggested that the systematic search by modern advertising writers for the most pleasing words and images in the language has had the indirect effect of exhausting and cheapening the vocabulary of pleasure and approval, with the result that it has become well-nigh unusable for the purpose of serious poetry or literature. Thus poets and writers have been left with the vocabulary and imagery of gloom, pessimism, and despair, as the main area of emotions still undespoiled by their commercial competitors.[47]

Some of the constraining effects of science on literature are paralleled by its effects on the fine arts. Here, too, we find an increasing constraint on rich ornamentation, and an increasing stress on clear or functional design. Considerations of the cost of ornamentation sometimes play a role, but often the less highly ornamented models are the more expensive, since they may require costlier materials. This "functional design" often turns out to be a design stressing and exhibiting one particular function of the building or object designed, or one particular theme to be emphasized in the work of art. This implies some difficulties where multifunctional designs or solutions are required. Sometimes, of course, ambiguity or multiplicity of meaning can itself be made a major function of the poem or a work of art; in that case many other traditional aspects of a work of art are likely to become subordinate to that purpose. Perhaps it could be said that while clarity and ambiguity were often kept somehow in balance in many of the traditional styles of art, both of them tend to become special functions under the impact of the example of science. Instead of seeking a balance between clarity and ambiguity, the modern artist may often find himself tempted to pursue either one or the other with the single-minded determination which we found characteristic of the scientific mind, and many works of art may thus tend to become either particularly clear, striking, and easily communicable, or else particularly complex, recondite, and difficult.

Even in the field of music some trace of this tendency can perhaps be found. We find on the one hand the demand for appealing, understandable, or even "singable" tunes in modern compositions, and on the other hand the development towards increasing technique and increasing craftsmanship for a small circle of fellow connoisseurs. During the earlier, classic age of science in the eighteenth century, its constraining effects on music were of a different nature, issuing in a stress on the importance of classic models and of fixed rules of composition, such as the rules for counterpoint and fugue, somewhat similar to the stress on classical models in literature and poetry earlier in the eighteenth century. Thus we learn that the young Beethoven was criticized for departing in certain of his works from the classical sonata form.

As these few examples indicate, none of these constraints was necessarily decisive or insuperable. Beethoven proceeded to make his own rules as he went on to his great compositions, and to some extent creative artists in all fields have tended to follow his example. We hear so much of the material benefits which science has brought to civilization, and of the great fears which it has aroused in many minds, that there may be some merit in a more sober study of the many limited but real contributions and constraints which science has brought and is still bringing to our culture.

Scientific Influences in Cultural and Social Conflicts and Decisions

If it were possible to trace the influence of humanistic civilization on the outcome—or at least on the temporary or intermediate outcome—of various scientific conflicts or decisions, it might also be rewarding to investigate the effects of science on the outcome of social or cultural conflicts or decisions, and thus on the general course of humanistic civilization. In the field of economic theory and policy, it might thus be possible to assess the impact of mathematical thinking, in the form, for instance, of Keynesian economics, on public opinion, as well as on the attitudes of governments towards such problems as inflation and the fiscal policies designed to balance or reduce the fluctuations of the business cycle. The long-standing search for a "scientific tariff," or the more recent search for scientific strategies in economic development in underdeveloped areas are perhaps other examples of attempts to apply scientific methods to economic practices with more or less success.

In industrial relations, attempts to apply some of the findings or methods of science to relationships of capital and labor have become associated with the phrase "efficiency engineering" and the name of Frederick W. Taylor; they have also been evident in more recent efforts to replace the emphasis on such efficiency engineering by the more modern emphasis on group psychology and human relations programs. Attempts to devise scientific measures for the output of workers or for the productivity of labor for an entire industry, or even for an entire country, have had an increasing influence in recent years on the drawing up of wage agreements, such as in the automobile industry in the United States, and even on entire national economic policies. In business life, psychology as well as sociological research methods have been applied to competitive advertising, as well as to management-employee relations; and spokesmen of management have appeared increasingly concerned about some science of organization or science of management which might permit our increasingly huge governments or private corporations to administer their own affairs and to determine their own behavior with greater effectiveness.

In the field of cultural and social decision, science has greatly increased the power of parents to decide when and where their children should be born. Through the development of techniques for marriage counselling it has begun to influence decisions concerning marriage; and through the spread of vocational counselling it is having an even greater influence on individual choices of occupation.

The development of aptitude and intelligence tests has at the same time had an increasing influence on decisions by organizations and institutions, as well as by society at large, as to which individuals are to be given the greatest opportunities for sharing in leadership, for becoming members of the so-called *élite* or for filling otherwise highly specialized social roles. In the field of race relations and racial conflicts, decisions once made on the basis of popular folk-lore or spurious race theories are increasingly modified in the light of the results of more serious investigation of the facts.

In political conflicts, decisions can be anticipated, at least to some extent, with the help of new sampling techniques and survey methods, such as the Gallup Poll and more recent refinements in public opinion research. At the same time, other attempts to apply the findings of social science to political conflicts, or to the decision of major problems of political organization, are still in their infancy. Science has, of course, been applied to warfare with such vast results as to threaten the physical survival of most of the world's large cities in the event of another world war; and the application of science to methods of anti-atomic defense, or to dispersal schemes for cities and industries, or finally to problems of international disarmament and inspection under the auspices of the United Nations, has done little as yet to reduce the danger.

Yet here again it seems possible that these last and seemingly most inhuman triumphs of science in the military field may turn out to have given a new impetus to some of the most vital forces of our humanistic culture. Sir Winston Churchill, on November 3, 1953, told the House of Commons: "It may be that . . . when the advance of destructive weapons enables everyone to kill everybody else, nobody will want to kill anyone at all."[48] By thus transforming warfare so that it no longer resembles what it had been from the stone age until the middle of the 1940's, and by depriving war of any positively desirable end for the nominal victor, science now may force mankind to reappraise its humanistic heritage and to reconstruct some of the fundamental habits and institutions of humanistic civilization. In any such thinking and reappraisal, the methods and results of science itself would no doubt be helpful; but it seems at least equally certain that the vividness, the multiplicity, the intuitive sophistication, and the presentational vitality of the humanistic approach would be no less essential. If civilization is to save itself by transcending the potentially suicidal aspects of its heritage, it will have to use for its self-preservation all the resources of scientific, humanistic, and religious thinking, which through their creative interplay have always been the sources of its growth.

IV. THE JOINT ENTERPRISE OF GROWTH

The problems of growth and progress have occupied thoughtful investigators in the sciences and the humanities for a long time, and in fields ranging all the way from the study of biological evolution to the history of civilizations and to the psychology of individuals. There seems to be fairly wide agreement that increased bigness is not necessarily growth, and that growth perhaps cannot be measured by any single indicator. Certain of the many dimensions of growth, however, stand out. In its relationship to its environment, an organization, organism, or culture may be said to grow if it shows an increasing independence from a larger range of different environments; if it shows increasing power to produce changes in its environment without having to accept correspondingly large changes in its own structure; and if it shows increasing openness towards its environment in the sense of an increasing capacity for deriving information and resources from it. The addition of new channels for absorbing information, similar in a way to the addition of new senses or sense organs, might thus represent a major dimension of growth.

With regard to its inner arrangements a system or organization may be said to grow if it shows increased articulation, that is, an increased precision and combinatorial richness in the rearrangements of its internal elements. Usually, growth seems to be accompanied by a greater rise in the volume of internal transactions—that is, activities within the system—than in the volume of external transactions—that is, transactions occurring across the system's boundaries. Some aspects of this lag in the number of external transactions compared to the number of internal ones is known to the designers of machinery as the "scale effect";[49] it is known to an historian of civilization as "the transference of the field of action" from the external to the internal realm,[50] and to traffic engineers as the "traffic problem."[51] By the same reasoning, we may expect growth to imply a transference of problems from the outside to the inside of the growing system or organization, and thus at least a tendency towards the increase of internal conflicts, potential deadlocks, and challenges.[52]

Growth and Strategic Simplification

The increasing complexity of the range of external environment dealt with, on the one hand, and the possibly even more rapidly increasing internal complexity of the growing thing itself, on the other, are offset and compensated at least in part, in cases of successful growth by what might be called "strategic simplifications." Growth in organizations and progress in technique often appear to imply just such a simplification of some crucial link or coupling in the chain of interlocking and self-sustaining processes by which the organization is kept going. Thus the maintenance of an ever growing written tradition is facilitated by the invention of increasingly simple alphabets and increasingly simple methods of writing and, eventually, printing. The many tasks of modern languages are facilitated by the sloughing off of many of the ancient inflections and the replacement of their semantic functions by means of word position and context. Other examples of this process of strategic simplification are the replacement of trolley tracks by rubber tires; of telegraph wires by radio; and many others ranging all the way to the symbol structure of the central theories of physics, increasing in simplicity from the days of Ptolemy and Copernicus to the more general formulations of Newton and Einstein.[53]

In none of the cases of simplification cited by Toynbee do we find a simplification—or a reduction in the number of elements—for any of the systems cited as a whole, be they systems of transportation, communication, or theoretical physics. On the contrary, each system as a whole is becoming more complex; it is particular crucial or strategic links in it which are becoming simpler. Thus a modern radio station is a far more complicated piece of electric equipment than the original wire telegraph of Samuel Morse; but its ability to transmit signals without telegraph wires permitted man to put its increased complex of resources to other and more fruitful uses. With this qualification, however, there seems to be a good deal of evidence supporting Toynbee's surmise that some such strategic or crucial simplifications may well be essential for any extended process of growth.

An important special case of such strategic simplifications might perhaps be seen in the replacement of gross operations or experiments with major physical resources by much simpler and quicker operations or experiments by means of symbols. An increasing shift from operations with gross resources towards a growing proportion of operations by means of symbols is thus perhaps another criterion of growth; and most of the cases of what Toynbee has called "etherialization" as an important aspect of growth could be brought perhaps under the heading of this and the preceding paragraph.[54]

Growth and Increasing Self-Determination

Growth, it follows from the foregoing, may be conceived as an increase in self-determination, or, in more engineering terms, as an increase in autonomous or self-steering performance. Self-steering or autonomy requires the receipt of information from the outside world, as well as from the behavior of the system itself, and in particular the feeding back of information regarding the system's own past behavior into the determination of its future action. A more extended degree of autonomy or self-determination requires, moreover, the use of information stored in the memory of the organization, or system, or individual. Autonomous decisions are then made when data recalled from this internal memory are brought to bear, in the light of current data received from the present environment. An increase in the richness and diversity of relevant data from the environment, on the one hand, and an increase in the richness and diversity of such past data stored in memory, on the other, are both necessary, though not sufficient, conditions for an increase in self-determination, and thus for major growth. Other characteristics of an increased self-determination could be expressed as responsiveness, adaptability, effectiveness of recognition of relevant data in terms of their partial correspondence to other data stored from the past, and initiative through the recombination and reapplication of new patterns derived from the rearrangement of associated parts of old patterns of information recalled from memory.

Growth and the Changing of Goals

All these aspects of self-determination have some bearing on the criterion for economic growth suggested by Simon Kuznets: the ability of an economic system or organization to reach or at least approach whatever goals it happens to have set for itself.[55] Growth, however, need not be confined to the ability to reach chosen or given goals. Rather, the capacity to choose new goals, and the range of potentially available goals for such choice, might in itself be a significant dimension of growth. Thus the growing system must be a system with increasing self-determination; it must be a system with a capacity to seek goals; and, for even greater growth, it must be a system or organization which shows the capacity to change goals and to transform itself by restructuring its own inner arrangements.

The capacity to seek a given goal is usually improved through a capacity for *learning*, that is, a capacity for some limited internal structural rearrangement leading to a change in subsequent responses to repeated or similar external stimuli. Goal-changing is in a sense a more thoroughgoing change in internal structure, and implies a greater capacity to learn. Learning capacity, in turn, can be related to the proportion of relatively uncommitted inner resources, that is, of inner resources which are at least available for relatively rapid recommitment from their present uses; and the capacity to transform oneself or one's own inner structure or, in religious language, to be "born again," might represent in some sense the highest form of the capacity to learn.

What could all these considerations mean for the growth of our civilization? Perhaps they might mean this. If our civilization should continue to grow, we may well expect that it will become simpler in many of its crucial aspects; subtler in the ranges, shadings, details, and discriminations of the processes with which it can deal; more inward in the wealth of its memories and mutual internal transactions, and more open in the number and effectiveness of its contacts and intake channels for information from the outside world and the whole universe; more self-controlled in its behavior and more childlike in its capacity to learn; more powerful in its ability to change its environment while preserving its own essential patterns undisturbed, and more prone to use symbols rather than gross power in order to foresee more exactly the consequences of its actions; more resourceful, and more willing to distrust the adequacy of its own resources; more ready to seek and encounter the grace of new vital and relevant information from the vast unknown universe outside itself; and more ready to be born again in the remaking of its inner structure in response to new external or internal challenges.

If our civilization should seek this path of growth, it might well pursue the search for greater simplification and openness, better self-determination and greater combinatorial richness and precision, by the methods of science. But already for everyone of these dimensions just mentioned, the sensitivity and awareness of the characteristically humanistic vision would have indispensable contributions to make. For other aspects of the openness to the unexpected, the ability to recognize subtleties, and the ability to respond and to be born again, the humanistic and religious approach would be essential. More than ever in the past of our civilization, the enterprise of growth would require the interplay of all the faculties of the human mind and heart.

Growth and Motivation

Throughout this discussion we may seem to have stressed the cognitive rather than the normative aspects of both art and science. Yet we are persuaded that the knowledge of that which ought to be cannot be divorced from the knowledge of that which is, and that the effort to know what is good cannot be separated from knowing what is real. For we call "good" that which is not self-destructive in its consequences, either for the mind of the doer, or for the smaller or larger community in which he lives. In its crudest terms, to be good means not to commit suicide, and not even to commit it by instalments. In religious language, it means to seek life everlasting, to love God—the vast reality around us—and to love our neighbor as ourselves. In the language of Kant's philosophy, it means so to act that the motive of our actions could be a general law of nature, or so to act that the principle of our action could be a general principle of human society; or finally, so to act that no human being shall be treated by our action as a mere instrument but as an end withal.

Each of these descriptions of good or ethical behavior implies the necessity of a lively and responsible concern for the consequences of one's actions. A man's attempted estimate of the probable consequences of his own act helps us distinguish the good intention of a doctor (even if his operation should not succeed) from the callous recklessness of the quack (even if by some great good luck his patient should survive). Good will, in Kant's sense, thus necessarily implies the will to know as far as possible the probable consequences of one's acts.

Yet if we know what is good (in the sense of knowing what behavior would probably have good consequences), we may not necessarily be motivated to do it. There are minds that can be aimed but are not loaded: on their triggering to action nothing follows but a click. Science has no direct influence on human motivation. It depends on the existence of human motives which it does not create, although its results may help to strengthen them. The humanities, on the contrary, present us with configurations of symbols and experiences that may have profound effects on the configuration of our motives. What is in science at best a by-product, the effect on the social, moral, and spiritual motives of human beings, can become—though it need not become—one of the central concerns of the humanities.

To be sure, the humanities can try to avoid this challenge. In the short novel *Tonio Kröger*, the young Thomas Mann suggested that the artist must be like an actor or like a dead man, for he must portray emotions he must not share. "Sincerity" and "good intentions" were thus left to the bumbling efforts of well-meaning amateurs. But the more mature Thomas Mann brought his great novel *The Magic Mountain* to the opposite conclusion; his hero risks his life, in dubious battle, in order to share the fate, the hopes, and the good intentions of his fellows. The inescapable burden of cognition, the inescapable responsibility for making one's battle less dubious in value, and one's acts and intentions less precarious in outcome—none of these should deflect our attention from the core and foundation of both scientific and humanistic work: the deep and self-renewing motivation of men and women to compassionate, merciful, and competent action.

NOTES

1. Susanne K. Langer, *Philosophy in a New Key* (New York: New American Library, 1951), pp. 63–82. Cf. also the same author's distinction between "discursive" and "non-discursive" forms in *Feeling and Form: A Theory of Art Developed from Philosophy in a New Key* (New York: Scribner, 1953), pp. 29–32, 39–41, 371–80, 385–7, 393–7.

2. This distinction between the scientist's and the artist's focus of attention does not tell us how men look upon problems of ethics and morality; they could do so as scientists or as artists. Examples of both approaches can be found in the history of thought, either in sharp separation, as in the contrast between the ethics of Jeremy Bentham and of Friedrich Nietzsche, or in conjunction, as in the comedies of Molière or in the writings of Rousseau.

3. James B. Conant, *Science and Common Sense* (New Haven: Yale University Press, 1951), pp. 37–41.

4. *Ibid.*, p. 38.

5. Demonstration by Professor Ashley Montagu, September, 1953.

6. For the former concept, see Julian S. Huxley, *Evolution in Action* (New York: Harper, 1953), pp. 124–76. For the concept of economic growth, see particularly Joseph A. Schumpeter, "Theoretical Problems of Economic Growth," *The Journal of Economic History*, Supplement VII, 1947, pp. 1–9; and Simon Kuznets, "Measurement of Economic Growth," *ibid.*, pp. 10–34.

7. Cf. Norbert Wiener, *Cybernetics; or, Control and Communication in the Animal and the Machine* (Cambridge, Mass., and New York: Technology Press and John Wiley, 1948), pp. 189–91.

8. Cf. H. T. Pledge, *Science since 1500* (London: H.M. Stationery Office, 1939).

9. Max Planck, *Scientific Autobiography, and Other Papers* (New York: Philosophical Library, 1949), pp. 33–4.

10. W. B. Cannon, *The Way of an Investigator: A Scientist's Experiences in Medical Research* (New York: W. W. Norton & Co., 1945); J. S. Hadamard, *An Essay on the Psychology of Invention in the Mathematical Field* (Princeton: Princeton University Press, 1945).

11. G. Kepes, *The New Landscape in Art and Science* (Chicago: P. Theobald, 1956).

12. Cf. Godfrey and Monica Wilson, *The Analysis of Social Change* (Cambridge: Cambridge University Press, 1945).

13. Le Corbusier's sketch showing certain proportions of Notre Dame Cathedral (Paris), Le Corbusier exhibit, Massachusetts Institute of Technology, 1952; and Le Corbusier (Ch. E. Jeanneret-Gris), *New World of Space* (New York: Reynal & Hitchcock, 1948), p. 23, and *Le Modulor: Essai sur une mesure harmonique à l'échelle humaine applicable à l'architecture et à la mécanique* (Boulogne: Editions de l'Architecture d'Aujourd'hui, 1950).

14. Cf. Charles Singer, *A Short History of Science to the Nineteenth Century* (Oxford: Clarendon Press, 1941).

15. Cf. Norbert Wiener, *Extrapolation, Interpolation and Smoothing of Stationary Times Series* (Cambridge: Massachusetts Institute of Technology Press, 1949), pp. 9, 21–3, 70–1.

16. Cf. Karl Jaspers, *Von der Wahrheit* (Munich: Piper, 1947), pp. 37–42, and *Tragedy is not Enough* (Boston: Beacon Press, 1952), pp. 14–17.

17. Cf. Ross Ashby, "Can a Mechanical Chess Player Out-Play its Designer?" *British Journal for the Philosophy of Science*, III, No. 9 (May, 1952), 44–57.

18. Karl Jaspers, *Tragedy is not Enough*, pp. 43–4.

19. James B. Conant, *On Understanding Science* (New Haven: Yale University Press, 1947), *passim*; cf. the "principle of fecundity" in S. K. Langer, *Feeling and Form*, pp. 8–9.

20. On this point see also Hermann Weyl, *Philosophy of Mathematics and Natural Science* (Princeton: Princeton University Press, 1949), pp. 155–6.

21. Jacques Hadamard, *An Essay on the Psychology of Invention in the Mathematical Field*; Walter B. Cannon, *The Way of an Investigator.*

22. Eve Curie, *Madame Curie* (Garden City: Doubleday, Doran and Co., 1937).

23. Philipp Frank, *Einstein: His Life and Times* (New York: Alfred A. Knopf, 1947), pp. 85–9; see also Max Brod, *Tycho Brahe's Weg zu Gott* (Leipzig: Wolff Verlag, 1916), pp. 114, 125–7, etc.

24. John Von Neumann, "The Mathematician," in Robert B. Heywood, ed., *The Works of the Mind* (Chicago: University of Chicago Press, 1947), pp. 180–96.

25. George D. Birkhoff, *Aesthetic Measure* (Cambridge: Harvard University Press, 1933). For other discussions of the problem of simplicity or economy of means in art or science see also H. Weyl, *Philosophy of Mathematics*, pp. 146, 147, 155–8, 191; L. L. Thurstone, "Factor Analysis," in Philip P. Wiener, ed., *Readings in Philosophy of Science* (New York: Charles Scribner's Sons, 1953), p. 194; Philipp Frank, *Modern Science and Its Philosophy* (Cambridge: Harvard University Press, 1949), pp. 209–10; K. W. Deutsch, "On Communication Models in the Social Sciences," *Public Opinion Quarterly*, XVI, No. 3 (1952), 362–3.

26. James B. Conant, *Science and Common Sense*, and *On Understanding Science*.

27. Aldous Huxley, *Science, Liberty and Peace* (New York: Harper Bros., 1946), p. 1; Leo N. Tolstoi, "Preface to Carpenter's Article 'Modern Science,'" *Complete Works*, Leo Wiener, trans. (Boston: Dana Estes & Co., 1905), XXIII, 113.

28. P. W. Bridgman, quoted in John E. Burchard, ed., *Mid-Century* (Cambridge: Technology Press, 1950), p. 230.

29. Eliot D. Chapple and C. S. Coon, *Principles of Anthropology* (New York: Henry Holt & Co., 1942), pp. 360–1; L. K. Frank, Foreword in "Teleological Mechanisms," *Annals of the New York Academy of Medicine*, L (Oct., 1948), 189–96; and Herbert Simon, "Model Construction in the Social Sciences," in Paul Lazarsfeld, ed., *Mathematical Thinking in the Social Sciences* (Glencoe, Ill.: Free Press, 1954), pp. 398–400.

30. Some of the problems are discussed in Margaret Mead, ed., *Cultural Patterns and Technical Change* (UNESCO, 1953).

31. For instances of discussions of this kind, see the *Bulletin of Atomic Scientists*, *passim*; George F. Kennan, *American Diplomacy, 1900–1950* (Chicago: University of Chicago Press, 1951); Thomas I. Cook and Malcolm Moos, "The American Idea of International Interest," *American Political Science Review*, XLVII, No. 1 (March, 1953), 28–44.

32. Such a process is described by G. Bachelard, *Les Intuitions atomistiques* (Paris: Boivin, 1933).

33. It is also characteristic that at first this spread of bookkeeping had the opposite effect; a decree of 1299 in Florence prohibited the use of the new Arabic numerals because of the ease with which they could be fraudulently altered. In the long run this consideration was outweighed by the need to handle continually increasing numbers, far beyond the convenience of the traditional combination of roman numerals and verbal expressions. The development is well illustrated in the works of François Rabelais; in the first two Books of *Gargantua and Pantagruel* (1532–4) his purposes are served by a combination of roman numerals and verbal forms (for 16,000 he writes *xvi mille*). But in the Third Book (1546), the debts of Panurge and the colonizing schemes of Pantagruel have to be described in millions and billions, and the new Hindu-Arabic numerals, the result of the efforts of mathematicians to systematize notation, achieve recognition in a major work of literature.

34. Cf. Eugen Rosenstock-Huessy, *Out of Revolution* (New York: Wm. Morrow & Co., 1938), pp. 547–50; reprinted in *The Driving Power of Western Civilization* (Boston: Beacon Press, 1950), pp. 95–8.
35. Cf. Charles Singer, *A Short History of Science*, pp. 5–93.
36. Cf. John Hermann Randall, Jr., *The Making of the Modern Mind* (2nd ed.; Boston: Houghton Mifflin Co., 1940), pp. 230.
37. On the possible bearing of some of these considerations on the relative technological stagnation of France towards the end of the nineteenth century, see David S. Landes, "French Entrepreneurship and Industrial Growth in the Nineteenth Century," *Journal of Economic History*, IX (May, 1949), 45–61.
38. Philipp Frank, "The Role of Authority in the Interpretation of Science," in L. Bryson *et al.*, eds., *Freedom and Authority in Our Time* (New York: Harper Bros., 1953), pp. 361–3.
39. See n. 13 above.
40. C. Day Lewis, *Short is the Time: Poems 1936–1943* (New York: Oxford University Press, 1945), p. 68; W. B. Yeats, *Collected Poems* (New York: The Macmillan Company, 1949), p. 154. The poem by D. H. Lawrence is reprinted in Selden Rodman, *An Anthology of Modern Poetry* (New York: Modern Library, 1938), pp. 114–15.
41. "Ballade des langues envieuses" in *The Great Testament*, French text and English prose translation in G. Atkinson, ed., *The Works of François Villon* (London: E. Partridge Ltd.–Scholartis Press, 1930), pp. 162–5, 263. Cf. also *The Poems of François Villon*, H. B. McCaskie, trans. (London: Cresset Press, 1946).
42. See n. 13 above.
43. Chicago: P. Theobald, 1956.
44. See, for instance, the well-known study by Richard Hofstadter, *Social Darwinism in American Thought, 1860–1915* (Philadelphia: University of Pennsylvania Press, 1944).
45. See David Riesman, *The Lonely Crowd* (New Haven: Yale University Press, 1950), and *Faces in the Crowd* (New Haven: Yale University Press, 1952).
46. Some of this hostility is described by Julien Benda, *La Tradition de l'Existentialisme* (Paris: Grasset, 1947). Cf. also S. Kierkegaard, *Fear and Trembling* (Princeton: Princeton University Press, 1952).
47. See S. I. Hayakawa, "Poetry and Advertising," in L. Bryson *et al.*, eds., *Approaches to Group Understanding*, Sixth Symposium, Conference on Science, Philosophy and Religion (New York: Harper Bros., 1947), pp. 369–74.
48. Prime Minister Sir Winston Churchill, Speech to House of Commons, *New York Times*, Nov. 4, 1953, p. 3:6.
49. J. K. Roberts and Edward L. Gordy, "Development" in C. Furnas, ed., *Research in Industry: Its Organization and Management* (New York: Van Nostrand, 1948), pp. 22–5.
50. A. J. Toynbee, *A Study of History* (2nd ed.; New York: Oxford University Press, 1935), III, 192–217.
51. *Ibid.*, pp. 209–12.
52. Cf. J. S. Huxley, *Man Stands Alone* (New York: Harper Bros., 1941), pp. 1–33; Toynbee, *A Study of History*, III, 195–204, 212–16.
53. *Ibid.*, III, 174–91. Cf. also the references to H. Weyl and L. L. Thurstone in n. 25 above.
54. Toynbee, *A Study of History*, III, 174–92.
55. S. Kuznets, "Measurement of Economic Growth," *The Journal of Economic History*, Supplement VII, 1947, pp. 10–34, esp. pp. 24–7.

II

"Those Scattered Rays Convergent": Science and Imagination in English Literature

by

F. E. L. PRIESTLEY

ART, SCIENCE AND PERCEPTION; THE ROYAL SOCIETY AND THE POETS; SWIFT AND ADDISON; WORDSWORTH AND SHELLEY; BROWNING AND TENNYSON; ROBERT BRIDGES AND AFTER

The two great mental disciplines which we distinguish by the names scientific and humanistic are by nature not inevitably antagonistic; on the contrary, each is to a considerable extent impoverished by alienation from the other. The proper kind of harmonious and fruitful relation can be established between them only when there is a broad and sympathetic understanding of the nature of each discipline as an activity of the human mind, its aims, its methods, and the special qualities of its achievements. Further, there must be a recognition that both science and the humanistic pursuits are of fundamental importance as different, but equally valid, parts of man's intellectual and spiritual adventure; their validity differs in kind, not in degree.

To establish this sort of understanding and recognition is by no means easy, and it can certainly not be made easy by simple definitions of "what science is," or "what art is." Simple definitions have, in fact, been a main source of misunderstanding. It is folly to expect vast, complex, and varied products of man's mental activity carried on through long periods of time to be reducible to a neat formula; the result can be only a distortion or at best a fragmentary truth. Nevertheless, although it is wise to be wary of definitions, it is necessary to understanding to make efforts at discrimination, tentative generalizations. Defining the boundaries is a vast and meticulous task, but a rough topographical survey, based on limited exploration, may be a useful preliminary.

The following paragraphs offer an exploration of this sort, obviously limited and incomplete, but not, I hope, useless. They start with some suggestions about the nature and aims of science and the humanities; they then turn to historical illustration to suggest by implication something of the nature, aims, and methods of literature, and also something of the conditions under which science becomes a valuable source of material for the poet. First, then, to the more general treatment.

I

One often encounters nowadays the view, put forward with neat formality by Martin Johnson in *Science and the Meanings of Truth*, that science, art, and moral philosophy are all mental activities of an analogous nature: Dr. Johnson extends to the other realms of thought the doctrine of Eddington, that in science we discover not things, but structures of relations. All human thought is an attempt to discover in, or impose upon, the flux of experience a structure of relations which will give to the flux a comprehensible form. In so far as the structure satisfies the criteria of coherence and communicability, it is true. The kind of coherence and the kind of communicability, like the kind of structure, are not the same in each area of activity (or the different areas of experience with which each activity deals), so that instead of speaking of the meaning of truth, Dr. Johnson speaks of the *meanings*. He by no means settles the difficult problems he raises, particularly of the nature of coherence and of communicability, but his work has the great virtue of recognizing the possibility of variety of "*meaning*." He is also careful to avoid giving a special value to one kind of meaning; each kind is valid within an area of experience. This is a welcome change from the *esprit simpliste* which would reduce all validity to one mode, which restricts truth to "empirical verification," and accepts only one sort of explanation as significant. (Works by Carnap, A. J. Ayer, and Wittgenstein contain illustrative examples of this sort of simplification.)

But while good as far as it goes, Johnson's approach does not go far enough. It has the disadvantage, first, of chopping experience up into separate areas, marked out by different kinds of experience, ordered to different ends, and apprehended by different kinds of mental process. It is important, to be sure, to recognize these differences, but there is a danger in over-emphasizing them. The Baconian tradition (as I shall note below) established not only a separation but often an antithesis between science and poetry, between thinking and feeling. This too often leads to a popular view of the human mind as capable of only one activity at a time; the scientist reasons without feeling and without flights of imagination; the poet feels and imagines but cannot think; and so on. The truth of the matter, that the same kind of human mind engages in all these activities, and that a poet like Dante, Shakespeare, Milton, or Goethe thinks, feels, and imagines all at the same time, tends to be forgotten. This is a truth of which the humanist is constantly reminded by the objects of his study; music, art, poetry, philosophy, the works of the great literary historians, all remind him that powers of thought, imagination, and emotional apprehension are constantly co-operative and co-existent in the creations of the human mind. This is, of course, equally true of the works of great scientists.

Nor is it sufficient to describe art merely as the ordering of experience. For one thing, art has a very special way, or rather many very special ways, of ordering it. We may grant that order, or form, is essential to art, that the form must be what Roger Fry calls "significant form," and that the material so ordered is the material of experience. But it is only necessary to bring together a tragedy like *King Lear*, and a comedy like *Volpone* or *L'Avare*, an epic like the *Iliad*, a mock-epic like *The Rape of the Lock*, a ballad, a song of Burns, a masque, paintings by Titian, Poussin, El Greco, Canaletto, Turner, and Matisse, and set against each other musical compositions by Couperin le grand, Mozart, Debussy, and Stravinski, to recognize the innumerable variety of forms which any art can select, and the innumerable variety of kinds of significance those forms can convey. The truth is, of course, that every art includes a vast number of *genres*, each of which may develop its own special form or forms, and its own kinds of significance. There are thus several kinds of tragic form, several kinds of tragic effect; within the one artist's work, for example, *Hamlet, Romeo and Juliet, Macbeth*, and *King Lear* all differ fundamentally in form and effect. In fact, while it is possible to define *genres* in broad terms, each work of art tends to create its own form and its own effect.

Art, then, although it represents, like science, an attempt at significant ordering of experience, differs in that each work of art offers a unique significance. In so far as the work of art belongs to a *genre*, it will of course gain in meaning; knowledge of the general nature of tragedy gained from a wide acquaintance with classical and French classical tragedy tends to illuminate the special qualities of Shakespeare's, and familiarity with Shakespeare's tragedies will enrich the understanding of the general nature of tragedy. The work of art grows in meaning in the context of other art; *genre* illuminates *genre*: mock-epic helps realize epic, comedy tragedy and tragedy comedy; epic, drama, and poetry help reveal the novel, the novel and drama the epic, and so on. But the ultimate uniqueness of the vision of the single work of art remains. Art as a result tends constantly to illustrate the manifold richness and variety of human experience. It resists the tendency of structure to simplify. The humanist, who produces or enjoys the work of art, is likewise concerned with the complexity of human nature and of the human situation. As humanist, he finds it difficult to think of man as a mere complex of biological urges, endocrine secretions, or conditioned reflexes. He accepts these reductions by generalization, in which the individual human being disappears, with full recognition of their usefulness, but with firm reservations. He has read the literature, heard the music, and seen the graphic art produced by men; he is aware of the epic grandeur and heroism, of the misery and debasement of man; of the astringent wit, ebullient humor, dry satire, pathos, anger, and affection with which man has viewed himself and his world, of the unearthly beauty and grotesque ugliness. He has learned to see the human situation and humanity itself in a multiplicity of aspects, and to accept its complexity and richness.

Science and art, then, differ fundamentally in their mode of ordering experience, as Whitehead long ago recognized in his contrast of the abstract and the concrete with which scientist and poet are concerned (*Science and the Modern World*, 1926, chap. v). As he says elsewhere (*The Concept of Nature*, 1920, p. 163), "the aim of science is to seek the simplest explanations of complex facts." The poet's aim is to overcome the tendency of language to reduce complex experience to a succession of simple elements; Coleridge gives to the poetic imagination the function of recreating a whole experience from the parts into which reason analyzes it. Browning presents the problem of the poet very clearly in his description of Sordello's inability to "give the crowd perception,"

> Because perceptions whole, like that he sought
> To clothe, reject so pure a work of thought
> As language; thought may take perception's place
> But hardly co-exist in any case,
> Being its mere presentment—of the whole
> By parts, the simultaneous and the sole
> By the successive and the many.

And again,

> '. . . within his soul,
> Perception brooded unexpressed and whole.'

An important aim of art, and more particularly of literature, is to convey, with the aid of aesthetic pattern, the complex quality of a whole experience; the revelation of that quality may in itself present the human significance of the experience, or the artist may use further means to present his conception of the significance. The relation between "direct" poetry (or "poetry of statement") and "oblique" poetry is by no means simple; by a poet like Tennyson, for example, statement is often used in an introductory mode, presenting the experience in simple terms as preparation for a more complex and subtle treatment through symbol; this is his characteristic technique in *In Memoriam*, as the frequent comments on the limitations of merely discursive language make clear.

Art is at once expression and communication; if critical theory has at times emphasized the one rather than the other, it has never completely ignored either. The eighteenth century (and to a great extent the nineteenth) paid chief attention to the public function of the artist, but recognized also very sharply the differences in vision of Pope and Dryden, Tennyson and Browning; if the twentieth century has swung to another extreme, it still expects the poet to publish, and still asks what he has to say to his readers. As a matter of fact, a study of the history of literary criticism reveals a firm continuity and stability; variations in emphasis, drawing out of implications, and clarifications of ambiguities leave the main structure not only intact but well buttressed. It is as illuminating now to read Dryden or Johnson or Hazlitt on Shakespeare as any modern critic, and Aristotle's *Poetics* is still the foundation and the fresh source of general dramatic criticism.

It is also significant that the fluctuations of emphasis, or changes in fashion, of criticism have little effect on the judgment of great art. There is, for example, a singular agreement among critics of Homer and of Shakespeare in both neo-classical and Romantic periods; they recognize and value the same qualities, even if they also find other qualities. Again, Dryden is able to see the essential qualities of Chaucer, just as Pope sees those of Donne, or Arnold those of Byron, despite great changes in language, verse technique, and social milieu.

All this suggests that the classical theory of the universality of art, of the power of art to speak to all men of all times, has a sure foundation in the facts of human experience. The popular notion of the subjectivity and relativity of art and of aesthetic theory has, on the other hand, a foundation in fallacies. The first of these fallacies is, of course, the naïve separation and opposition of subjective and objective. The second fallacy arises from a misunderstanding of an aspect of the very universality of art: if art is addressed to all men, what it says to each is of the same significance, and every man is his own infallible critic. This fallacy has the flattering effect of providing an aesthetic theory for the man who knows nothing about art but knows what he likes. Johnson, to be sure, announced that final judgment in literature rested with the common reader; it is important to remember that he said "reader" and to realize what he meant by the word. It is obvious that art can only communicate to those trained to recognize and discriminate the uses of the medium: the opinions of even Gene Tunney on Shakespeare are less valuable than Dover Wilson's.

There is, of course, a subjective element, and an important one, both in the work of art as it is produced, and in its effect, since the experience of both artist and audience is made up of elements some of which are unique, some common. The meaning of a work of art to me will consist partly of the relation it establishes with private elements in my experience, partly of the relation with elements I know I share with most, partly of the relation with elements I do not recognize in my own experience, but know of in others. What art communicates, then, is neither simply subjective nor objective; it is certain that great art communicates in large part meanings of experience which nearly all recognize.

Precisely what and how art communicates is a problem to which I should be rash indeed to offer a solution. I hope it has already been made clear that what art is trying to convey is the complex quality of experience, and that its media are also complex. It follows that there can be no simple statement of its operations. The term "communication" is in itself an illustration of the problem, since it at once suggests a false simplicity of means and of substance. It is impossible to teach poetry for many years without becoming more and more aware, for example, of the part played by rhythm and tonality in conveying meaning, and I do *not* mean in simply arousing an emotion or mood in the reader or auditor. Those whose work is not with language take an extremely simple view of the nature of language, of the way it conveys meaning, and the kind of meaning it conveys. Those whose lifetime is spent working with language, and particularly with the subtle and complex language of poetry, realize best the fallacy of the attempt to reduce meaning to "propositions." The meaning of "meaning" is not reducible to an easy formula.

Another difficulty I propose only to suggest is that of defining the effect of aesthetic pattern itself. In part this effect is the "aesthetic emotion" of pleasure aroused by symmetry, balance, repetition, and so on, but the effect in the lyric of "open" or "closed" patterns goes obviously beyond this into a realm of suggested meaning. Music perhaps here offers a parallel. One thing seems certain, that the aesthetic effect is not purely emotional.

Finally, there is the problem of the didactic element in literature. It is this that the unliterary see as the "communication"; they find in the *Ancient Mariner* a mnemonic version of an SPCA tract. Such prosaic and utilitarian interpretations of "didactic" may constitute a comment on our notion of education; they at least reflect an eagerness to reduce literature to an applied art. The real didacticism of literature, and perhaps of other art, is, as I have suggested, contained in its apprehension of the quality of experience. This apprehension is by no means to be equated with the acquiring of moral maxims, or propositions from the copy-book, of laws of psychology, or of the rules of good citizenship. Some of these may be acquired from the study of literature, as from the study of life; but it is not the chief business of literature to furnish them.

To quote Browning's *Sordello* again on the function of the poets:

> The office of ourselves,—nor blind nor dumb,
> And seeing somewhat of man's state,—has been,
> For the worst of us, to say they so have seen;
> For the better, what it was they saw; the best
> Impart the gift of seeing to the rest.

Poetry is, as Arnold said, a criticism of life; it leads toward self-knowledge, towards knowledge of man. Literature assists man towards the humanistic ideal, to "see life steadily and see it whole." It aims at the cultivation of the whole of that many-sided creature, man; at development of all his powers of awareness, sensitivity, judgment. Its motto must be, "Nihil humanum a me alienum puto."

It necessarily follows that literature, and the humanist who learns from it, cannot afford to ignore science, which represents one of man's most vast and most successful areas of endeavor, and which itself constitutes one of the most important modes man has created for viewing himself and his world. This does not mean that literature can or should adopt the methods of science; on the contrary, the arts must, to be significant, preserve their own aims and methods. The subject-matter, the language, the forms, the kinds of meaning sought in science and in literature are not the same, and should not be; the symbols used by each have only the name in common. The area at which science and poetry co-operate, or at which literature is enriched by science, may be suggested by an excursion into the history of English literature in its relation to science.

II

It will be convenient to follow convention by taking the foundation of the Royal Society as the starting point; it does in fact mark in an official way the beginning of new movements in both science and literature after the Puritan Interregnum. And under the patronage of Charles II both at first flourished together in an atmosphere of harmony, and Dryden is at once a chief architect of neoclassicism in literature and a herald of the new age of science. His work reveals an effort to reconcile old and new. He praises modern achievement in a well-known passage of his *Essay of Dramatick Poesie* (1668, 1684), in his verses to Dr. Charleton (1663), and in the fine apostrophe to the Royal Society in *Annus Mirabilis* (1667); at the same time he is by no means willing to join in wholly rejecting the past. As Neander in the *Essay of Dramatick Poesie* he acts as mediator, and indeed the whole aim of the *Essay* is to present the arguments for both sides in the controversies about ancient and modern, French and English, rhymed and unrhymed drama. Dryden's own poetic practice enforces the conciliatory approach; he draws his imagery from a wide range: from science old and new, from the traditional Aristotelian philosophy or from the new, from classical literature, from commerce, from the techniques of crafts, from popular proverbs, from popular "natural history," from sports. The part played in his verse by any single one of the kinds of imagery is limited; his early poetry, in which he is closest to the "metaphysical" tradition, makes freest use of scientific ideas, but they are never as dominant as in more strictly "metaphysical" poets. There is, in fact, a tendency for the new criteria of clarity, simplicity, and directness in writing to lead him away from "metaphysical" practices towards a choice of fewer, simpler, and more familiar images. Nevertheless, his earlier verse in particular offers good illustrations of poetic use of scientific ideas (as in stanzas 2, 3, 4, 17, 53, 161 ff. of *Annus Mirabilis*) and they occur occasionally even in later works (see *The Hind and the Panther*, Part III, ll. 366 ff. and 497 ff.). And the new style he and his fellows establish in both poetry and critical prose is not unrelated to the stylistic requirements of the new Royal Society.

The association, through the Royal Society, of science with the monarchy and, by implication, with the Established Church served to break the damaging associations of the Commonwealth period, when certain Puritan groups had strenuously espoused the Baconian philosophy and had antagonized the universities by attempts to turn them into schools of practical science. As R. F. Jones has observed ("Background of the Attack on Science in the Age of Pope," *Pope and his Contemporaries: Essays presented to George Sherburn*, 1949, p. 99), "The favor shown the scientists by His Majesty is no inconsiderable fact in the history of science, for it discouraged the critics of Baconianism from calling attention too openly to the Puritan past of the organization now sponsored by the King." (See also R. F. Jones, *Ancients and Moderns*, 1936, chap. v, for the Puritan educational aims.) The desire of the new Society to break the old link between science and Puritanism, and to modify the Commonwealth antagonism between new science and old learning is revealed further by the efforts of Thomas Sprat, in his *History of the Royal Society* (1667) to demonstrate that "Experiments will not injure Education," that "Experiments are not dangerous to the Christian Religion," and that "Experiments are not dangerous to the Church of England."

The nature of the membership of the Royal Society undoubtedly assisted the task of harmony; however much the presence of amateurs, of men of fashion, of distinguished figure-heads may have hampered the scientific work, it had the useful and important effect of bringing the activities of the scientist into a central arena of total national activity, and of creating a body of men joined by innumerable links to the main national institutions. The inclusion of the Archbishops of Canterbury and York and many of the bishops, and the absence of Hobbes, successfully protected the young movement from charges of either Puritan or atheist tendencies; the presence of men from the universities promised at least some degree of co-operation. One thing appears certain: however incomplete the harmony established by the Royal Society (and it was of course not complete), it does at least seem to have prevented a sharp division into hostile camps, and it is significant that two of Newton's champions and popular exponents are the great classical scholar, Bentley, and the Cambridge metaphysician, Samuel Clarke. It is significant of the effort towards harmony, too, that they both lectured in the series endowed by Boyle.

That the harmony was not complete depended on a number of factors. One was the implicit opposition between the tendency of literature to build on the past, and the feeling of scientists that science was breaking completely with former ages. This conflict emerged openly in the quarrel of ancients and moderns, the Battle of the Books. The kind of armistice in which the battle terminated perpetuated the separation of the combatants; the ancients were granted suzerainty in the arts, the moderns in science. Furthermore, the opposition was accentuated by the introduction of political issues. After 1688 the Whigs were increasingly inclined to adopt the doctrine of progress, and to identify themselves with the new age and the new science, as much Whig panegyric verse shows. The court still nominally patronized science through the Royal Society, but literature could look for support only to party; the Hanoverian attitude towards the arts is enshrined in Pope's *Epistle to Augustus*. As party strife became more acrid, the opposition poets turned more and more to satire, and to attacks on the corruption of the Whig party machine, which the Whigs defended by panegyrics on the growth of commerce, wealth, and science. The political conflict has nevertheless a limited effect on the conflict between science and literature, since poets and scientists are not each of a single party.

Another potential source of separation lay in the growing strength of the middle class. From Elizabethan times at least there are two traditions in English literature, a major one: aristocratic, Anglican, and classical in education; and a minor: bourgeois, Dissenting, and modernist in education. Bunyan, Defoe, and Richardson may be taken to represent the minor tradition; Dryden, Addison, and Fielding the major. The tendency of the minor tradition is to be seen in Defoe's *Tour of Great Britain*, with its emphasis on agriculture and industry and its ignoring of the beauties of scenery and architecture; it is the tour of a practical man with an eye to business. The education offered in the new schools for Nonconformists excluded by the Test Act from Oxford and Cambridge was also practical; these Dissenting academies increasingly stressed the study of modern languages and science. The growing influence of Dissenters in the Royal Society in the latter half of the eighteenth century would suggest that official science was allied by that time to the non-classical tradition. There is perhaps significance in the continuance at Cambridge, under men like Samuel Vince, of the purely mathematical Newtonian tradition, while the Royal Society had become largely concerned with the more empirical sciences.

III

The foregoing suggestions are offered simply as such; they are not intended to do more than indicate some of the possible elements in a complex development. For fuller treatments readers are referred to the works of R. F. Jones, cited above, and of Robert K. Merton. Even the facts of opposition between science and literature in the eighteenth century have by no means been fully ascertained: the grounds for such an opposition are certainly present but the extent of developed opposition is not so certain. Throughout the first half of the eighteenth century there is a considerable body of literary Newtonianism, as Miss Nicolson and others have demonstrated; poets like Prior and Thomson make extensive use of science. What gives the opposition to science an appearance of great strength is perhaps the towering figure of Swift, whose Grand Academy of Lagado constitutes at least the most vigorous comment on science in English literature of the period. But a close examination of Swift's actual position suggests that it may not be simple. If we leave aside the problem of Swift's motives (which demands a knowledge of Swift I make no pretense to) and consider merely *The Battle of the Books, A Tale of a Tub*, and *Gulliver's Travels*, we can perhaps draw one or two conclusions from the text. The first is that Swift is by no means ignorant of science, and is not expressing the contempt for science of the ignorant *littérateur*. The knowledge of astronomy which Professor Gould has shown Swift to possess (in "Gulliver and the Moons of Mars," *Journal of the History of Ideas*, VI [January 1945], 91–101) is not the sort to be casually acquired for purposes of satire by a literary man with no real interest in the subject. Gulliver's remark that the Laputans "were indeed excellent in two sciences for which I have great esteem, and wherein I am not unversed" need not be ironic; it seems rather a direct emergence of the Swift who understood the mathematics of astronomy, and who knew enough about music to build in Dublin one of the best choirs in Europe.

The second conclusion emerges from the recognition of a strongly unified theme throughout Swift's comments on science. In *The Battle of the Books* he remarks of the Moderns that "they have in speculation a wonderful agility, and conceive nothing too high for them to mount, but in reducing to practice discover a mighty pressure about their posteriors and their heels." He characterizes the pursuits of Gresham College, in the introduction to *A Tale of a Tub*, as the seeking of husks. The minds of the Laputans are "so taken up with intense speculations" that they are incompetent in all affairs; "their houses are very ill built . . . and this defect ariseth from the contempt they bear to practical geometry, which they despise as vulgar and mechanic. . . . They are very bad reasoners. . . . Imagination, fancy, and invention, they are wholly strangers to . . . the whole compass of their thoughts and mind being shut up within the two forementioned sciences (mathematics and music)." They are "so abstracted and involved in speculation, that I never met with such disagreeable companions." "His Majesty discovered not the least curiosity to enquire into the laws, government, history, religion, or manners of the countries where I had been, but confined his questions to the state of mathematics. . . ." By contrast, the idealized King of Brobdingnag takes a lively and intelligent interest in all matters of concern to the welfare of his people. The learning of the Brobdingnagians Gulliver describes ironically as "very defective, consisting only in morality, history, poetry, and mathematics, wherein they must be allowed to excel." The "defectiveness" of this education is indicated: mathematics in Brobdingnag "is wholly applied to what may be useful in life, to the improvement of agriculture, and all mechanical arts; so that among us it would be little esteemed. And as to ideas, entities, abstractions, and transcendentals, I could never drive the least conception into their heads." Speculation not closely related to practice, then, forms no part of Brobdingnagian pursuits; the ideal expressed by the kind is "that whoever could make two ears of corn or two blades of grass to grow upon a spot of ground where only one grew before, would deserve better of mankind, and do more essential service to his country than the whole race of politicians put together." And to "politicians" Swift would add "projectors" of all sorts.

It is in this context that the Academy of Projectors at Lagado belongs. Swift prepares us for the visit to the Academy through Gulliver's encounter with Lord Munodi (in chapter IV of Book III). Munodi embodies the ideal expressed by the King of Brobdingnag; his estates are well-farmed and prosperous, with "vineyards, corn-grounds, and meadows." Elsewhere in Balnibarbi, Gulliver notes, "except in some very few places I could not discover one ear of corn or blade of grass." Munodi's well-managed lands and well-built houses are threatened by the political power of the projectors, who "with a very little smattering in mathematics, but full of volatile spirits" have "schemes of putting all arts, sciences, languages, and mechanics upon a new foot." Where the projectors have power, "the whole country lies miserably waste, the houses in ruins, and the people without food or clothes." Munodi's own water-mill has already been destroyed to make way for a mill high on the mountain; "after employing a hundred men for two years, the work miscarried, the projectors went off," and the countryside now has no mill.

The projects pursued at the Academy work variations on the theme. Each proceeds from a fantastic hypothesis, and is directed either towards a fantastic end or a fantastic means of attaining an ordinary end. Much attention is drawn to the expense of the projects, and the number uselessly employed in them; outside the vast Academy with its five hundred rooms lie the wasted lands and miserable hovels of the people.

The implications seem uniformly clear. Swift is objecting to two things: the cultivation of a science divorced from human needs; and the cultivation of science to the exclusion of other studies related to human society. The words of Bacon seem applicable, in which he presents science as "a rich storehouse, for the glory of the Creator, and the relief of man's estate."

> This is that which will indeed dignify and exalt knowledge [says Bacon], if contemplation and action may be more nearly and straitly conjoined and united together than they have been; a conjunction like unto that of the two highest planets, Saturn, the planet of rest and contemplation, and Jupiter the planet of civil society and action. . . . Neither is my meaning . . . to leave natural philosophy aside, and to apply knowledge only to manners and policy. . . . The end ought to be, from both philosophies to separate and reject vain speculations and whatsoever is empty and void, and to preserve and augment whatsoever is solid and fruitful. (*Works*, VII, 105–6, quoted by F. H. Anderson, *The Philosophy of Francis Bacon*, 1948, p. 96.)

Swift would seem to be pursuing the same end, since he attacks "whatsoever is empty and void" not only in science, but in literary scholarship, political philosophy, and religion. This is not to suggest, of course, that Swift is a follower of Bacon, although it will be noted that Bacon gets off lightly in *The Battle of the Books*. But he and

Bacon agree at many points, perhaps at those where they are both Aristotelians.

Swift's treatment of Newtonianism is also not without significance. Although he repeatedly pokes fun at Epicurus and Descartes, he makes only one reference to "attraction demonstrated from mathematical principles." Aristotle's ghost, called up for Gulliver in Glubbdubdrib, predicts that it, like other systems of nature, would pass out of fashion. When one recalls the mockery and parody of Newton's views by other opponents, Swift's seems unusually restrained. All it asserts, after all, is that the Newtonian system is not the final and eternal system. Elsewhere in *Gulliver's Travels*, Swift makes fun of those who live in "continual disquietudes, never enjoying a minute's peace of mind" from fear of changes in the celestial bodies. The nearest of these apprehended dangers is the approach of the next comet, "one and thirty years hence." "They are so perpetually alarmed . . . that they can neither sleep quietly in their beds, nor have any relish for the common pleasures or amusements of life." Swift's comments make it clear that his satire is directed here at the folly of the fear, rather than at the predictions of the scientists, since the annihilation of the sun, and the contraction of the earth's orbit, are offered as remote events. At any rate, to sacrifice all relish for life through fear of the possibility of losing it typifies the kind of irrational misuse of life Swift decries; it is to become a "sort of struldbrug."

Swift's satire as a whole suggests some final implications: that the unity of knowledge is threatened, and that the true purpose of knowledge is being frustrated in many areas of thought. Pedantry in literary scholarship brings the same narrowness and dehumanizing as are found in Laputa. Controversies of Big-endians and Little-endians, of Tramecksan and Slamecksan, lead religion and politics away from genuine concern with human welfare and the amelioration of man's lot. Swift's satiric argument for breadth and humanity of outlook possibly suggests already the separation of science, literature, politics, and religion, or at least signs of such a separation.

It will perhaps throw some light on Swift's attitude in these matters if we turn for a moment to his contemporary, Addison. At many points Addison offers a contrast: he is not so strongly marked by genius, his emotions are very completely under control, he is a good Whig, and, although a thoroughly competent classicist and exponent of neo-classicism, has a strong interest in and enthusiasm for the new science. At Oxford Addison acquired "a sanguine estimate of the Cartesians, and approval of the mechanical experiments of Boyle and of the scientific spirit" (Peter Smithers, *Joseph Addison*, 1954, p. 19), and in 1693 delivered at the annual Encaenia commemorating the founders a Latin oration, "Nova philosophia veteri praeferenda est." He read much in natural history, was an enthusiastic Newtonian (as several papers in the *Tatler* and *Spectator* attest), and a disciple of Locke. Nevertheless, as Smithers shows (p. 17), his attitude was often very similar to Swift's, not only towards science, but towards literary scholarship:

> For the minutiae of scholarship Addison had a contempt which sometimes spilled over upon genuine projects of learning, for example upon "editors, commentators, interpreters, scholiasts, and critics, and in short all men of deep learning without common sense." (*Tatler*, 158.) It was for the dignity of a comprehensive culture that he reserved his admiration. . . . He was as suspicious of a devotion to detail in natural science as in scholarship. When carried to any length it outraged his sense of proportion: "There are some men whose heads are so oddly turned . . . that though they are utter strangers to the common occurrences of life, they are able to discover the sex of a cockle, or describe the generation of a mite, in all its circumstances. . . ." (*Tatler*, 216.)

As Smithers, his latest (and first complete) biographer, points out, this contempt for the student of minutiae leads Addison to ridicule those in the Royal Society who seemed to him to be "preoccupied with unimportant hair-splitting unbecoming to a rational man" (p. 250).

The general theme of Pope's *Dunciad*, and particularly of Book IV, offers another close parallel. The triumph of Night, the return of Chaos, is brought about by the "poring Scholiasts,"

> Wits, who, like owls, see only in the dark,
> A Lumber-house of books in ev'ry head,
> For ever reading, never to be read,

and by their naturalist counterparts, of whom the Goddess of Dulness remarks:

> Yet by some object ev'ry brain is stirr'd;
> The dull may waken to a humming-bird;
> The most recluse, discreetly open'd, find
> Congenial matter in the Cockle-kind;
> The mind, in Metaphysics at a loss,

May wander in a wilderness of Moss. . . .
O! would the Sons of Men but think their Eyes
And Reason giv'n them but to study Flies!
See Nature in some partial narrow shape,
And let the Author of the whole escape:
Learn but to trifle. . . .

Pope's concern, like Addison's and Swift's, is that narrow specialization will destroy the more comprehensive view. In part, their position is the common neo-classical one, which sets up the ideal of the *honnête homme*, and which, in its more rigorous French form, discouraged professionalism generally (as in the attempt to prove Plato and Aristotle amateurs); the extreme case in England is provided by Congreve, whose carelessness towards literature so shocked Voltaire. But there is much more in it than the separation of "gentlemen" and "players"; there is a genuine ideal of balanced and harmonious development of the human powers in the individual, and of a comprehensive and harmonious society, both of which are threatened by the cultivation of intellectual "splinter groups" with special interests.

One real source of opposition between scientist and humanist is touched on in *Gulliver's Travels*. An important part of the third voyage of Gulliver deals with language. In one section Swift parodies the naïve notion that language consists merely of arbitrary symbols for material objects, or should so consist (a notion popularized in our own time by Hayakawa). In another section, through the device of the language-frame, he parodies all mechanical notions of language and of composition, and probably also attacks the Epicurean "fortuitous concourse of atoms." The narrowing of the function of language by the Baconian tradition is part of a general narrowing, a reduction of reason to calculation, an opposing of reason and imagination, a limiting of truth to "scientific truth." Ultimately, there is in the Baconian tradition a contempt for the arts.

Although scientists like Ray and his successors are fully alive to the imaginative response to new marvels (see, for example, Ray's *The Wisdom of God manifested in the Works of Creation* (1691), and Derham's *Physico-Theology* (1713) and *Astro-Theology* (1715) —these are typical of a tradition), and to the validity and usefulness of imaginative contemplation, the pressure of Baconian psychology and aesthetic theory was strong. Addison's papers on the imagination make this overwhelmingly clear. Since Bacon reduces art to mere play, and the creation of untruth, he denies art any validity (except a restricted function as very simple allegory). It is significant that Addison, a major literary proponent of neo-classicism, accepts essentially the Baconian position, as A. S. P. Woodhouse has shown (in *English Studies Today*, edited by Wrenn and Bullough, Oxford, 1951). One effect in art of this critical position is to foster a completely literal representationalism—to give to the Aristotelian *mimesis* the simplest meaning of mere transcription. Another is to separate form and content, substance and medium, and to foster the theory of "applied decoration"—reason supplies the real content, fancy tacks on the trimmings. Erasmus Darwin's *The Loves of the Plants* may be taken as the orthodox product of these critical principles, or perhaps even Canning and Frere's *The Loves of the Triangles.*

IV

However, poets are fortunately not entirely submissive to psychological theories, and have continued to use fancy (or imagination) whenever something stimulated it. And new ideas usually did. The strong response in poetry and prose to the new cosmology and its implications for the religious view of nature, and to Newton's optics, have already been mentioned. Of the two, the latter has proved the more permanent, and it might be well to consider why. The appeal of Newton's cosmology to the poetic imagination lay largely in its implications. It provided some direct stimulus, of course, in its vision of massive bodies hurtling through immense space, held from flying off tangentially into chaos by the divine power; and in the thought of the vast and lonely orbits of the comets. But the dominant effect in literature of the *Principia*, particularly as popularly interpreted by Bentley, Clarke, Whiston, and Derham, was derived from a vision of order and harmony produced by the divine immanence. Since this was not a new vision, but one merely given new force and sanction by Newton's system, as the system became more familiar and commonplace, it lost its power to stimulate the imagination. The effect of the *Optics*, on the other hand, was more immediately dependent on the novelty of its own ideas. Not only were these fresh and stimulating in themselves, but they were susceptible of becoming permanent and useful symbols for the poet. (Gravitation, to be sure, had some success as a symbol, but it suffered from over-literalness of application.) For centuries poets had had a rich and complex symbolism of light and darkness, and of colors; Newton's theories allowed further enrichment and complexity, as Shelley's "dome of many-coloured glass" and Browning's wide use of prism and of burning-glass symbols illustrate. The *Principia* and the *Optics*, then, illustrate two different kinds of contribution science may make to poetry, and suggest which is the more permanent and valuable.

The revolt by the poets against the Baconian tradition in the latter part of the eighteenth century coincided with a strong resurgence of it in science. The dominance of Newtonian celestial mechanics, which had achieved in England a static orthodoxy from which the Continent was free, gave way to the newer, more empirical sciences, particularly those centered around electrical and chemical experiments. At the same time, the degree of caution and rigor imposed by the search for mathematical laws was abandoned, and the kind of sweeping hypothesis advanced tentatively by Newton was now freely advanced assertively. This opened the way for the expansion of empiricism on a truly imperial scale; science, like its patron saint, was prepared to take all knowledge for its province, and even plant its flag on Parnassus—after duly levelling the mountain. The process may be illustrated briefly by considering the tendency of Burke's *Inquiry into the Origin of our Ideas of the Sublime and Beautiful* (1756), the most important document in eighteenth-century empiricist aesthetics. Burke's whole approach is marked by a shift of attention from the quality of the work of art to the nature of its effect on the audience, and by a shift from explanation of this effect in terms of reason to one in terms entirely of emotion. The sublime thus becomes that which produces a certain kind of emotion. "Beauty is, for the greater part, some quality in bodies acting mechanically upon the human mind by the intervention of the senses." Note the quality in "bodies," not in works of art, and the "acting mechanically"; the principle of association, and a mechanical theory of sensation, lead to a mechanical aesthetic. The cow, with its rough coat and angular lines, becomes "sublime" to Gilpin, while the horse, smooth and smoothly curved, is "beautiful." The setting of aesthetic criteria in emotional response had its value as a counter to the arid formality of over-intellectualized aesthetics, but it had great dangers. The Gothic novel of terror was the logical companion of Burke's aesthetics. Moreover, if the value of a work of art lies merely in the emotion it rouses, no criterion in art can transcend the renowned "I know nothing about art, but I know what I like," and Bentham's dictum that "pushpin is as good as poetry, the quantity of pleasure being equal." It is notable that Bentham did not go on to say "pushpin is as good as political economy, or science, the quantity of pleasure being equal." The whole tendency of this empiricist movement was to reduce art to a pleasing but irrelevant pastime. It is this that the Romantics oppose. Their opposition takes various forms, but all involve a rejection of Bacon's theory of knowledge. Blake offers the simplest position, Coleridge the most subtle and complex. Keats contents himself with an assertion of the truth of the imagination's apprehensions; Wordsworth relates imagination to the intuitive, as distinct from the discursive, reason; Shelley adopts a thorough, if somewhat distorted, Platonic doctrine. The effect of the revolt and the reassertion of the validity of poetic truth is to reinvigorate poetry on the one hand, but to widen the gap between science and the humanities on the other. It must in fairness be admitted, I think, that science bears the greater responsibility for the gap. The poets do not deny the validity of science, nor do they ignore it (Wordsworth and Shelley are evidence enough here); it is the scientists who deny the validity of poetry and who ignore it.

One point of limited co-operation is established through David Hartley. In his *Observations on Man, his Frame, his Duty, and his Expectations* (1749), Hartley elaborated a system of psychophysical parallelism based upon a principle of association. From the last paragraph of Newton's *Principia* (second edition), the Queries in the *Optics*, and Newton's letter on the aether to Boyle, Hartley proceeds to ascribe to Newton a doctrine of neural communication through "vibrations" of a "very subtle spirit," or aether. Vibrations or "vibratiuncles" of the brain are invariably accompanied by (but are not recognizably the cause of) sensations in the mind. Patterns of vibrations are established by association, so that, vibrations A, B, and C having recurred together or consecutively in the past, henceforth vibration A, being repeated, will arouse also vibrations B and C. So also, sensation *a* will bring with it sensations *b* and *c*. On the physical side, Hartley's theory is close to that of Hobbes, substituting vibration of a fluid medium for contact of moving particles. In carefully separating the physical and mental, Hartley is, as he himself recognizes, close to Cartesian and Occasionalist theories. What gave Hartley's theory its special importance, however, was his account of how, through association, simple ideas of sensation move through six stages of complexity, to form ideas of Imagination, Ambition, Self-Interest, Sympathy, Theopathy, and the Moral Sense. Since each stage involves all of the lower stages, the system has the advantage of emphasizing a kind of unity in the mind's operation, of not setting sensation in opposition to the Moral Sense, for example. It also gives a function to ideas of Imagination in the production of ideas of Theopathy and the Moral Sense; and even though it dispenses with the faculty of Imagination it avoids the simple opposition of ideas of Imagination and "true" ideas.

It seemed for a time to offer the Romantic poets a useful scheme of their own operations. This is particularly true of Wordsworth, whose use of Hartley's system is most extensive and various. The poem, "Influence of Natural Objects in calling forth and strengthening the imagination in boyhood and early youth," suggests how the poet tends to find in Hartley at least part of the explanation of his own development:

> . . . thus from my first dawn
> Of childhood didst thou interwine for me
> The passions that build up our human soul;
> Not with the mean and vulgar works of Man;
> But with high objects, with enduring things,
> With life and nature; purifying thus
> The elements of feeling and of thought,
> And sanctifying by such discipline
> Both pain and fear,—until we recognize
> A grandeur in the beatings of the heart.

The effect of associations formed of sensations received from the wrong sort of natural object is described in "Ruth":

> The wind, the tempest roaring high,
> The tumult of a tropic sky,
> Might well be dangerous food
> For him, a Youth to whom was given
> So much of earth—so much of heaven,
> And such impetuous blood.
>
> Whatever in those climes he found
> Irregular in sight or sound
> Did to his mind impart
> A kindred impulse, seemed allied
> To his own powers, and justified
> The workings of his heart.

Nor less, to feed voluptuous thought,
The beauteous forms of nature wrought,
Fair trees and gorgeous flowers;
The breezes their own languor lent;
The stars had feelings, which they sent
Into those favoured bowers.

Yet, in his worst pursuits I ween
That sometimes there did intervene
Pure hopes of high intent:
For passions linked to forms so fair
And stately needs must have their share
Of noble sentiment.

Behind these passages, one is aware of the Hartleyan scheme, relating sensations from natural objects to complex "ideas" of Imagination, Ambition, Self-Interest, Sympathy, and the Moral Sense; it is in describing the formation of character that Wordsworth relies most directly upon Hartley.

He also finds Hartley of use in explaining stages in the development of his own response to nature. This is clearly seen in the familiar "Lines composed a few miles above Tintern Abbey," when he describes how the "beauteous forms" of the natural scene have affected him during years of absence.

I have owed to them
In hours of Weariness, sensations sweet,
Felt in the blood, and felt along the heart;
And passing even into my purer mind,
With tranquil restoration. . . .

The repetition of "felt" and the careful succession of "blood," "heart," "mind" are here significant. Wordsworth goes on to recount the change that has come with the years in his response to nature. In his boyhood, nature was to him "all in all."

The sounding cataract
Haunted me like a passion: the tall rock,
The mountain, and the deep and gloomy wood,
Their colours and their forms, were then to me
An appetite; a feeling and a love,
That had no need of a remoter charm,
By thought supplied, nor any interest
Unborrowed from the eye.—That time is past. . . .
For I have learned
To look on nature, not as in the hour
Of thoughtless youth; but hearing often-times

> The still, sad music of humanity,
> Nor harsh nor grating, though of ample power
> To chasten and subdue.

The sense of the beauty of nature is by no means lost, but is now enriched with human associations; nature still retains its ennobling function, and the poet recognizes

> In nature and the language of the sense
> The anchor of my purest thoughts, the nurse,
> The guide, the guardian of my heart, and soul
> Of all my moral being.

The Hartleyan doctrine is here serving less as explanation than as a mode of description; it performs much the same function as the Platonic myth Wordsworth uses to express the same theme in the "Ode on Intimations of Immortality." As a doctrine, associationism is already inadequate to Wordsworth; he is content to use its patterns formally. Its fundamental inadequacy is that common to all mechanistic psychology in the eighteenth century; it presents a passive mind acted on by the dead external object. As a poet, Wordsworth is aware of his own creative activity; he also needs to convey his sense of being in the presence of a living and active nature. Even in these early poems, then, the use he can make of Hartley is limited, and he supplements it by other structural elements. Philosophically, there is perhaps a conflict between associationism and Wordsworth's doctrine of an animated nature; poetically there is none: together they convey very exactly the response of the child to nature and the response of the thoughtful man, contemplating human suffering against the quiet beauty of its setting.

Once one recognizes the part association plays in Wordsworth's poetry (as he himself recognized it with Hartley's aid), the richer meanings become clear. The theme of a simple poem like "Nutting" (later included in *The Prelude*) is seen to be related to that of the "Immortality Ode" and "Tintern Abbey" lines, and the "Elegiac Stanzas." The idyllic description of the hazel copse, and of the unthinking boy who breaks it open to the sky, departing enriched but saddened, bears a close relationship to the secure world of beauty, broken into by the knowledge of sorrow. The figure of the boy becomes a complex symbol of humanity and also of time. The figures and situations in the *Lyrical Ballads* are similarly enriched through the close association of natural objects and human situations. Another plain example is given by Wordsworth at the end of the "Immortality Ode":

> The Clouds that gather round the setting sun
> Do take a sober colouring from an eye
> That hath kept watch o'er man's mortality;
> Another race hath been, and other palms are won.
> Thanks to the human heart by which we live,
> Thanks to its tenderness, its joys, and fears,
> To me the meanest flower that blows can give
> Thoughts that do often lie too deep for tears.

Although Wordsworth's own particular poetic interests and practice allowed him to make use of Hartley's theories, the general tendency already noted of any associationist psychology to put a limit on the creative activity of the poet made it ultimately of minor use, and the main strength of Romantic poetry came from other traditions, which even Wordsworth called on to supplement Hartley. Wordsworth is well beyond Hartley when he writes (*Prelude*, v):

> Visionary power
> Attends the motions of the viewless winds,
> Embodied in the mystery of words;
> There, darkness makes abode, and all the host
> Of shadowy things work endless changes,—there,
> As in a mansion like their proper home,
> Even forms and substances are circumfused
> By that transparent veil with light divine,
> And, through the turnings intricate of verse,
> Present themselves as objects recognized,
> In flashes, and with glory not their own.

Shelley, who had not Wordsworth's formal education in mathematics and astronomy, but who was an ardent amateur follower of the more empirical sciences of chemistry and electricity, makes a much fuller use of the ideas of science, as Professor Grabo's studies have shown. Wordsworth, apart from tributes to Newton, and some passages of cosmological reference which reflect the tradition of Cambridge Newtonianism, seems to make very little use of his scientific training. Shelley, however, uses scientific concepts very freely to supply major structural elements, as in the familiar "The Cloud," and for major symbolic patterns, as in *Prometheus Unbound*. He also shows some interesting uses of scientific imagery, and well illustrates the tendency already mentioned, for the new imagery to enrich old symbolism. It is perhaps worth examining the familiar passage from *Adonais* already referred to:

> The One remains, the Many change and pass;
> Heaven's light forever shines, Earth's shadows fly;
> Life, like a dome of many-coloured glass,
> Stains the white radiance of Eternity
> Until Death tramples it to fragments.

This is a beautifully complex passage. The theme of the One and the Many, stated in philosophical terms, itself presents the paradox of the relationship of Unity and Manifoldness, Permanence and Change. The word "Life," following on the calm and abstract statement, brings the paradox from the cosmological to the less abstract human level, where the antitheses of One and Many, the Remaining and the Changing, are applied to the human antithesis of Life and Death. The theme is worked out through a highly formal pattern of Life, Eternity, Death, powerfully aided by three linked symbols—the dome, the "white radiance," the "many-coloured glass." The dome is an ancient and complex symbol; its trampling by Death, along with its shape, here gives it the archetypal meaning of the egg, whose breaking is at once a destruction of a world and a release to new life (a similar meaning attaches to the chrysalis). The shattering "to fragments" of the "many-coloured glass" deliberately exploits the paradox; the reduction of Life to fragments (the many) reveals the One (the white radiance). The ancient symbolism of light, as the source of life, or symbol of spirit, and the equally ancient color-symbolism of white as purity (reinforced by "stains") has added to it here the Newtonian doctrine of color: the white light is the complete light; through the glass come various fragments, each a real part of the real radiance, but never giving the full aspect of it. The superb way in which Shelley here enriches the old symbolism by new theories of light seems a fulfilment of Sprat's hope that science would provide new images for poetry. This goes far beyond the mere recognition of possible new metaphors.

V

A few years after Shelley's death Tennyson and Browning, two of his great admirers, and to some extent his disciples, are making similar uses of science. Browning's education was highly informal, and mainly literary, musical, and artistic, but his symbols and images are drawn from a wide range of scientific knowledge, particularly from optics. The subtle and ironic symbol of the goldsmith's working of the ring in *The Ring and the Book* is more strictly technology than science, and the technology of art at that, but he repeatedly makes use of the symbol of the spectrum, of the reduced image from a condensing lens, and so on. These are of course by no means new, but it is true that Browning shows far less knowledge of, and far less concern with, contemporary science than Tennyson. The new sciences (which were already developed in Shelley's day, but which play little part in Shelley's poetry) are geological and biological, with the concept of evolution most significant. Browning is by no means well read in these sciences (a letter shows complete confusion of Darwin and Lamarck), but he has some important references to evolution.

Evolution in its general and pre-Darwinian sense is, in fact, a concept closely relevant to Browning's central pattern of thought, as its use in the early poem *Paracelsus* shows. The theme presented dramatically in this poem is one which Browning elaborates throughout his later works; Paracelsus has been seeking the secret of the universe, the total meaning of the cosmos, by an attempt at complete knowledge. The farther he pursues his quest, the more oppressed he becomes with a sense of failure, and the more inclined to doubt the validity of his instinct to seek knowledge. It is only on his death-bed, in a moment of vision, that he sees that the meaning lies in the very failure and incompleteness against which he rebelled; that the universe is not to be known as a complete, static scheme, since it is a dynamic process of becoming. The meaning of human life, too, lies in the process; his search had not led to the kind and degree of knowledge that he had expected, but it had indeed brought him wisdom—finite wisdom, to be sure, since he is a finite creature. As he now looks at the world, he sees its essentially dynamic nature, and sees too that the process is understandable in terms of its end. It is here that Browning's imagination seizes on the pattern of evolution, a pattern whose essence is dynamic (as opposed to the static Great Chain of Being), and which becomes intelligible as a process leading to the production of its highest form, man. The passage illustrates splendidly the enrichment of the poetic imagination by a scientific concept; naturally, the poet does not use the concept as a scientist would, and the accuracy of the concept as science is as irrelevant here as the accuracy of Milton's astronomy in *Paradise Lost.* The teleological emphasis in the passage might offend the scientist, but Browning is not relying on science for his teleology; for him the universe must have purpose and design or be entirely devoid of meaning. What he relies on is the primary intuition of purpose and value in life, the intuition which started Paracelsus on his quest and which he was wrong to doubt. The universal purpose is not demonstrable by science or any other mode of thought; it has to be a primary assumption that there is some point in living and striving. Once a purpose is assumed, fragments of plan can be seen, particularly in the order of science. The passage is too long to be quoted in its entirety; I quote with omissions, although this is to destroy the power and form of the speech:

> The centre-fire heaves underneath the earth,
> And the earth changes like a human face;
> The molten ore bursts up among the rocks,
> Winds into the stone's heart, outbranches bright
> In hidden mines, spots barren river-beds,
> Crumbles into fine sand where sunbeams bask—
> God joys therein. . . .
> When, in the solitary waste, strange groups
> Of young volcanos come up, cyclops-like,
> Staring together with their eyes on flame—
> God tastes a pleasure in their uncouth pride. . . .
> The grass grows bright, the boughs are swoln with blooms
> Like chrysalids impatient for the air,
> The shining dorrs are busy, beetles run
> Along the furrows, ants make their ado; . . .
> savage creatures seek
> Their loves in wood and plain—and God renews
> His ancient rapture. Thus he dwells in all,
> From life's minute beginnings, up at last
> To man—the consummation of this scheme
> Of being, the completion of this sphere
> Of life: whose attributes had here and there
> Been scattered o'er the visible world before,
> Asking to be combined, . . .
> Imperfect qualities throughout creation,
> Suggesting some one creature yet to make,
> Some point where all those scattered rays should meet
> Convergent in the faculties of man. . .
> And from the grand result
> A supplementary reflux of light
> Illustrates all the inferior grades, explains
> Each back step in the circle. . . .

At two points, Browning comes into conflict with the science (or perhaps rather the popular scientists) of his day. He objects often to their proneness to the genetic fallacy, the fallacy of explanation by origins (see, for example, his passage on evolution in *Parleying with Furini,* ix, x), and he objects to the simple opposition of fact and fiction, reason and fancy, which restricts all knowledge to "scientific" knowledge, and relegates to the artist the production of pretty falsehoods. For Browning, "fact" is a construct of the mind, a built relation, an interpretation, as is "fiction." He discusses this at length in *The Ring and the Book,* treats it ironically in the analogy of the making of the ring, and in *La Saisiaz,* where "fact" and "surmise" work an ironic interplay. He repeatedly tries to explain to his nineteenth-century audience the truth of art, the kind of truth art conveys, and its modes of conveying it.

Tennyson's concern with science meets Browning's at these points. He also finds it necessary to challenge the assumption that there is only one kind of truth, and one approach to it, but he uses quite different methods. Instead of a partly defiant, partly humorous irony, Tennyson tries a direct and serious expression. His problem is essentially one of terminology: "reason" and "rational" have been given a limited empirical denotation (as the "Rationalist Press" makes clear); terms like "faith" and "intuition" and "feeling" have similarly been narrowed. As a result, the full force of Tennyson's direct statements is often not realized, and the casual reader, leaping to the conclusion that the poet's use of "trust," "felt," and "faith," is as diluted as his own, finds the verse hesitant. In some poems (notably *The Ancient Sage*) Tennyson attacks the problem of knowledge directly and philosophically. More often, however, his real meaning is to be found in "oblique" passages, conveyed through symbols (as in sections XCV and CIII of *In Memoriam,* which are more central to the meaning of the poem than any of the discursive passages).

> 'For words, like Nature, half reveal
> And half conceal the Soul within.'

Tennyson's knowledge of science was wide and exact; we have Huxley's word for it that Tennyson knew as much science as any man in England. He was also well read in the classics, of course, and interested in a great variety of things, including Persian language and poetry. The sources of his imagery and symbolism are consequently various: classical myth, Oriental lore, history and legend, closely observed natural scenery, astronomy, geology, and "natural history." His use of scientific imagery is often brilliant; of the many examples two must suffice. The first is from *In Memoriam*. As the poet reflects on the theme of mutability, like Shelley he constructs a powerful pattern of antitheses.

> There rolls the deep where grew the tree,
> O earth, what changes hast thou seen!
> There where the long street roars hath been
> The stillness of the central sea.

As in Shelley, the symbols are ambivalent. The stillness of the central sea is in one aspect the stillness of death, contrasted with the roar of life in the long street. In another aspect the stillness of the central sea is the repose that has given place to the roar of change. The geological changes, which make the most permanent forms of earth, the everlasting hills and the sea, merely transitory, provide Tennyson in the next stanza with a brilliant symbol: with an eye which suddenly contracts the long span of geological time, he sees the successions of dissolving landscapes:

> The hills are shadows, and they flow
> From form to form, and nothing stands;
> They melt like mist, the solid lands,
> Like clouds they shape themselves and go.

The same sort of vision is present in

> The moanings of the homeless sea,
> The sound of streams that swift or slow
> Draw down Aeonian hills, and sow
> The dust of continents to be. . . .

My other example is an astronomical one, from *The Ancient Sage*. The Sage is replying to another poet, whose verses express the pessimistic naturalism made popular by FitzGerald's *Rubaiyat*. The Sage, while not minimizing the darker aspects of life, objects to the "black negation" which dwells on death to the denial of life, on sorrow to the denial of joy. He puts his whole position into four lines:

> But earth's dark forehead flings athwart the heavens
> Her shadow crown'd with stars—and yonder—out
> To northward—some that never set, but pass
> From sight and night to lose themselves in day.

The image of the great conical shadow of the earth racing through space, up which we peer to see the stars, and the sense of a universe of light outside that cone, are magnificently conceived and singularly apt as symbols.

Apart from scientific imagery, Tennyson's interest is however mainly in the implications of nineteenth-century scientific attitudes, and particularly in the ethical implications of evolutionary naturalism, which takes over the Utilitarian moral relativism and hedonism. Since this part of Tennyson's work is very familiar, and since it depends on implications relating to general thought rather than to poetry in particular, I say nothing of it here, although it is of the highest importance. Nor is there space here to discuss the response made to these implications by poets like Meredith and Hardy, later contemporaries of Tennyson, who adopt naturalism, each in his own way, as part of a general philosophy.

VI

Tennyson is popularly taken as the typical Victorian. Wrong as this judgment often is, in regard both to Tennyson and to what is typically Victorian, it is at least true that in his lively interest in science, in his awareness of conflict through the implications of science, and in his wish to find a compromise to retain the values and validity both of science and of other modes of thought, Tennyson stands for his age. In this sense, too, Robert Bridges is the last of the Victorians, and his *Testament of Beauty* is the testament of a great age. It would be easy to demonstrate the close relation of Bridges' poem to the main pattern of nineteenth-century thought; to show that the tensions expressed in Tennyson's *The Two Voices* and *In Memoriam*, and traceable through the main writers of the age, are here brought to a kind of equilibrium. Bridges' whole concern is, like Browning's, to reconcile antitheses. Just as Browning repeatedly rejects the popular oppositions of soul and body, head and heart, so on a more comprehensive philosophical scale Bridges attempts a vast assertion of the unity of experience. Differences should be recognized, but not allowed to cleave reality into an irreconcilable dualism. Bridges chooses as his symbol for the problem of dualism the Sphinx; "There is no pretence of hiding the unsolved riddle of life. The Sphinx lurks in all systems." The double nature of the Sphinx is a riddle, but also a fact; those systems of philosophy which proceed from a strict separation of spirit and matter, noumenon and phenomenon, reject the unity of the Sphinx.

The main direction in *The Testament of Beauty* is towards an assertion of unity. The popular oppositions are accepted as differentiations, but not as antitheses. Their range indicates the range of the poem: matter and mind; man and nature; reason and will; instinct and ethic; pleasure and duty; sensual beauty and the intellectual or spiritual; intellect and aesthetic or spiritual perception. All of these suggest the main problems with which the nineteenth century had been struggling; it is sufficient illustration perhaps to recall Huxley's *Man's Place in Nature* and then his *Evolution and Ethics*, or the controversy roused by Buchanan's *The Fleshly School of Poetry.*

The conceptual instrument Bridges uses throughout the poem is a variety of evolutionary doctrine, which allows him to view the whole of nature, including man, as a continuum, and yet to introduce a qualitative scale. In its general form, this is a continuation of suggestions made much earlier by Tennyson; what distinguishes Bridges' work is the thoroughness and detail of the working out, and the care to avoid letting qualitative differences develop into oppositions: man's "higher" faculties grow out of, and are constantly related to, his "lower" ones; the cultivation of the "higher" does not involve a denial of the "lower." All of man's qualities Bridges sees as more conscious development of qualities found in animals; he carefully relates conscious and unconscious thought, instinct and conscious ethical behavior, physical sense of well-being and intellectual and aesthetic pleasure.

His treatment involves extensive discussion of the relations of science and art, of ethics and art, of wisdom and beauty. Here again he is concerned with reconciling antagonisms. His whole poem is filled with scientific concepts, not only in the general framework, but in details of argument and in poetic details of imagery. It is itself a witness to the relation of science and aesthetics, intellect and sensibility, natural and spiritual beauty; it provides the example of its own philosophy. The jarring elements of nineteenth-century thought are here by the power of music wrought into a cosmos; as one reads the poem one is aware of the century of effort which it terminates, of the dissonances, partial resolutions, and suspensions which lead up to this final C major, as Browning would say. One is also aware that, as the poem was being written, a whole new set of themes was already dominant. Bridges was writing at a peculiar moment in the history of thought; for him and for many of his readers, science meant what it had meant to the Victorians: geology, biology, and the theory of evolution. But the focus of interest of science had already changed for the scientist, and was on the point of changing for the reading public, to physics, astronomy, and the theory of relativity. Bridges, with his great technical power and comprehensive mind, grasped a moment of stability which resolved and crystallized a century of thought; the new themes and new implications of science rapidly changed the perspective of his poem for the reader.

When one recalls that Bridges' *Testament* is the work of a poet roughly contemporary with Meredith, Hardy, and Charles M. Doughty, all of whom write poetry expressing general philosophies based to varying extents upon scientific ideas, and that Hardy's *Dynasts* and Doughty's *The Titans* are both of near-epic proportion and intention, one recognizes more clearly the special quality of their period. Since their time, new ideas both in science and in poetic theory have made the philosophical poem at once more difficult to read and less fashionable, and science more difficult to understand but perhaps no less fashionable. Curiously enough, the new science, heavily mathematical, highly subtle in its experimental techniques, and obscure in its implications, has taken a rather peculiar hold on the lay mind. It would perhaps be hard to match historically the variety of implications drawn (or stretched) from the theory of relativity, or from the Principle of Indeterminacy. General interest in science (theoretical as well as technological) reached in the 1930's and 1940's an amazing peak, if one may judge by the number and sales of works offering popular expositions. This meant that the poet could count on a measure of familiarity with certain concepts in his audience, particularly since the poetic audience is no longer the popular one of the last century, but a limited group, generally of the better educated. (I avoid the question of how far the modern poet has limited his own audience by his techniques, or how far his techniques are a response to the limited audience he expects.) It may well be that neither science nor poetry will ever again be popular in the earlier sense; it is perhaps significant that the modern poet very often expects the poetry-reading public to have also some knowledge of scientific ideas.

Thus, the Scots poet Hugh MacDiarmid makes great demands

upon his reader; he makes free and ingenious use not only of scientific ideas (mainly biological and physiological) but also of technical terms, often injecting his wry humor into a serious poem by deliberately mixing levels of diction. A good example is his "Ex-parte Statement on the Project of Cancer," in *Stony Limits* (1934). A somewhat simpler example from the same volume is "Thalamus," in which MacDiarmid uses anatomical details to support his theme.

> Busy as any man in those centres of the brain
> Where consciousness flourishes I yet cherish more
> The older, darker, less studied regions
> Of cranial anatomy through which, momently, pour
> Myriads of sensations, hundreds fleetingly combined into
> feelings, and now and then
> Among them a species of thought more profound
> Than any other that is known to man.
>
> The younger organs have not superseded
> These older ones which have a different mental life.
> I am aware at times between them
> Of co-operation and at times of strife,
> But these dark places intimate now and again
> A kind of knowledge the younger seldom recognize as such. . . .
> . . . But proud of their cortex few
> Have glimpsed the medial nuclei yet
> Of their Thalamus—that Everest in themselves
> Reason should have explored before it
> As the corpora geniculata before any star
> To know what and why men are. . . .

In poems of this sort, the scientific concept provides the main structure, as in some of the seventeenth-century "metaphysical" poetry. This is common practice with MacDiarmid, and a good many examples could be drawn from other modern poets (Alex Comfort's "The Atoll in the Mind" offers a close parallel).

Other poets make a more rapid, allusive use of scientific ideas. This is characteristic of early Eliot and Auden, and more recently of Christopher Fry, and could be abundantly illustrated from a great many contemporary poets. Fry makes particularly sophisticated use of allusion:

> We must be mellow,
> Remembering we've been on the earth two million years;
> Man and boy and Sterkfontein ape.
>
> We have a borrowed brilliance. At night
> Among the knots and clusters and corner boys

Of the sky, among asteroids and cepheids,
With Sirius, Mercury, and Canis Major,
Among nebulae and magellanic cloud,
You shine, Jessie. . . .
To take us separately is to stare
At mud; only together, at long range,
We coalesce in light.

Of course, the poetic imagination still responds directly to the scientific idea. Alfred Noyes, in what I think his best narrative verse, *The Torch-Bearers*, recounts with accuracy and a good sense of the dramatic, the stories of the great scientists; in this work he has also achieved what seems to me his best lyric:

Up-whispered by what Power,
Deeper than moon or sun,
Must each of the myriad atoms of this flower
To its own point of the coloured pattern run;

Each atom, from earth's gloom,
A clean sun-cluster driven
To make, at its bright goal, one grain of bloom,
Or fleck with rose one petal's edge in heaven?

What blind roots lifted up
This sacramental sign
Transmitting their dark food in this wild cup
Of glory, to what heavenly bread and wine?

What Music was concealed,
What Logos in this loam,
That the celestial Beauty here revealed
Should thus be struggling back to its lost home?

Whence was the radiant storm,
The still up-rushing song,
That built of formless earth this heavenly form,
Redeeming, with wild art, the world's blind wrong;

Unlocking everywhere
The Spirit's wintry prison,
And whispering from the grave, "Not here! Not here!
He is not dead. The Light you seek is risen!"

Or, to conclude, take a very different example, Walter de la Mare's use of the fact of phototropism to evoke all the rich, strange beauty of the moth:

Isled in the midnight air,
Musked with the dark's faint bloom,
Out into glooming and secret haunts
  The flame cries, 'Come!'

Lovely in dye and fan,
A-tremble in shimmering grace,
A moth from her winter swoon
  Uplifts her face:

Stares from her glamorous eyes;
Wafts her on plumes like mist;
In ecstasy swirls and sways
  To her strange tryst.

III

Tensions and Anxieties: Science and the Literary Culture of France

by
HARCOURT BROWN

THE HUMANIST BACKGROUND; THE AGE OF REASON AND RULES; DIVERSITY AND ITS FRUITS; FONTENELLE, VOLTAIRE AND POSITIVIST CURRENTS; THE SCIENCES OF MAN, POETS AND PSYCHOLOGY, RIMBAUD AND THE BAUDELAIRIAN BASIS, BIOLOGY AND THE NATURALISTIC NOVEL; TWENTIETH-CENTURY TENSIONS AND THE HUMAN DIMENSION; SAINT EXUPÉRY AND ALBERT CAMUS

Karl Deutsch has pointed out in his essay that there are many areas in which the opposing claims of scientists and humanists have led to intellectual and emotional tensions, some of them acute. Thus the creative imagination of a Romantic poet or novelist may be criticized in the name of scientific objectivity and precision; the wayward emotionalism of an Alfred de Musset or the bombast of a Victor Hugo may be replaced by the documentary flatness of naturalism and the sharp and uncompromising impassivity of the Parnassian poets; and in turn, science, under the name of scientism, comes to be cast out by a vigorous reaction towards a more sensitive and spontaneous art, a literary mode that seems less dogmatically a priori, more responsive to the ways in which human beings think they normally react. Such sequences can of course very usefully be presented and studied schematically, but it is clear that they are historical phenomena, with dynamic relationships. The present chapter will attempt to review one geographically and linguistically limited complex, an area of culture and creativity in which the diversity of causes and consequences may perhaps assist in achieving historical understanding.

I

It is commonplace to suggest that human society as we know it is inconceivable without all the arts, both fine and useful, without the impulse towards imaginative vision, towards humanism, as well as the strong impersonal desire to know, to satisfy objective curiosity. The most richly human figures of the past, Dante, Plato, Leonardo, Goethe, Shakespeare, Montaigne, Molière, have divided their attention between the two poles; they gave intuition, the poetic impulse, its full place, while recognizing that there is a realm in which the individual insight counts for less, the consensus for more, where

---

Some of the earlier pages of this essay are revised from an article published in *Diogenes* (Summer, 1953). A preliminary form of the whole discussion was read in a joint meeting of the History of Science Society and the Literature and Science group of the Modern Language Association in New York, December, 1948. The teachings and writings of George Sidney Brett, Professor of Philosophy in the University of Toronto, have contributed much to the point of view from which this admittedly incomplete account is written. To his memory, as to the good will and frank criticisms of many colleagues, both on and off the ACLS Committee, goes an expression of warm appreciation. A grant from the Penrose Fund of the American Philosophical Society has notably facilitated the collection of material for this and for other similar investigations.

what is known can be evaluated objectively, and its basis in experience repeatedly controlled. To allow logical analysis to create the illusion that experience may be broken into completely discrete parts is to betray the insight of every individual whose mind moves from the arts to abstract knowledge, even from enjoying the sunset to elementary arithmetic. And what is true for the individual mind as it moves through daily life is also true for humanity in general as its culture accumulates around it, its artistic product in museum and library, its science in the knowledge and skills of the engineer and the experimenter.

Thus it was only natural that in the sixteenth century the sciences should have received a strong and perhaps decisive directing impulse from the recovery and interpretation of the classic texts of Archimedes, Euclid, Hippocrates, Galen, and many others. While some of those who made the classical background of modern science available could be described as scientists, all of them were at the same time grammarians, translators, editors, scholarly publishers, whose accumulating experience and combined efforts gradually built up a consciousness of the tradition of the theory and philosophy of science, by putting important texts and commentaries in the hands of a large and omnivorous new reading public. Only gradually did experimental science become a discipline conscious of itself as something quite different from the older philosophic schools. The great figures of the seventeenth century, Descartes, Hobbes, Gassendi, Locke, Spinoza, Leibniz, still carry with them in their thought the imprint of philosophy as a mistress, unifying all knowledge.

Among these, however, some do more than others to create the independence of science, to identify it as a new field within which men could work without regard for the prejudices of the general public or the traditional positions of the schools. In this light, Gassendi is not as clearly a new man as Descartes, and Hobbes is less strikingly original than Leibniz or Spinoza. Similarly, after Locke and Newton, the scientists of the new national academies of the last third of the century are able to leave much of the traditional philosophic baggage behind them, and get on with their investigations in the new frame of reference. With the eighteenth century, theology loses its importance in the method of men who no longer look for final causes, while the poet and the artist, rejecting the necessity of the mathematical view, the outlook that "casts all in doubt," as John Donne had put it, still cling to a diction and a substance that long familiarity has hallowed. Thus the stage is set for an action full of strife, for a development of tensions and anxieties not exactly paralleled in all the earlier history of the race.

II

Central in the movement of ideas, a key to the situation in European thought as in politics, France exhibits and reflects vividly most of the impulses and tensions of the age. The intellectual atmosphere of Paris favored science as well as literature, promoted freedom of thought and speculation, enriched the imagination with new visions of truth, endowed the mind with qualities of clarity and precision, promoted humanism, and developed the sense of diversified values. In contrast with what was occurring elsewhere, in England and Italy, for instance, in France there was a unique and fruitful relationship between the creative literary imagination and the empirical and rational adventure of the scientific investigator. Objectivity, precision of argument, moderate conclusions, elegance of presentation, respect for accepted postulates, freshness of outlook and treatment, regard for the dimensions of the human form and for normal human usefulness, these are qualities which mark both science and literature in seventeenth-century France. To the vastness of *Paradise Lost* and the *Principia*, France opposes *Le Misanthrope* and the *Discours de la méthode*. While Frenchmen wrote works corresponding in scope to the large vision of the English authors, *Les Tragiques* of Agrippa d'Aubigné, *La Recherche de la vérité* of Nicolas Malebranche, such productions exhibit neither the genius that sustains the thought of Milton and Newton, nor the brilliant economy that marks the creations of Molière and Descartes. Yet all these works have an underlying kinship of method and outlook, a background of ideas and sentiments, instinctive attitudes, taste and prejudice, which in spite of conscious differences of direction and purpose, form and style, make one feel that here is a strong and homogeneous culture, present wherever the French mind is at work.

There is much evidence to corroborate the view that the society of France in this period was characterized by homogeneity. The diaries and correspondence of the time, published and unpublished, show that a foreign visitor could move across the country, from town to town, finding officers of government, clergy, merchants, lawyers and judges, Huguenot as well as Catholic, who offered no barriers to the exchange of views, who took an enlightened and unprejudiced interest in experience and ideas, discussing readily the problems and issues of their time and place. The records show that men of many nationalities did just that: Danes, Englishmen, Netherlanders, Italians, Scots, Swiss, Germans—none of them met a spiritual boundary that could not be crossed, a curtain, iron or silk, to stop the commerce in thoughtful interpretation and the recording of investigation and observation. At no time in this age do distinctions of class or wealth seem seriously to have impeded the formation of friendships, groups, and associations among persons from all parts of France who to our eyes make strange, even incongruous, company. A glance at the membership of the numerous *conférences* and *cabinets* of the age shows that the gap between Protestant and Catholic, even among the clergy, between merchant and professional classes, between the *homme d'épée* and the *roturier*, was not important when ideas were concerned. The fabric of society was an assured thing, not subject to change or upheaval; confident of its permanence, the Frenchman of the seventeenth century could let ideas come and go. Free discussion in a very wide area of thought was not deemed likely to lead to modification of the basic patterns of civil society.

Even a brief glance at this intellectual world, so foreign, perhaps even utopian to us, shows its features, its physiognomy, its reflection of a human reality. The tensions we know between scientist and humanist were not visible then; it is probable that we tend to read our own defects of confidence into the history of that age. The scraps of early physics we retain from high-school days hardly prepare us for the adventure of the young men of 1648 running long glass tubes up the façade of the town house of the Pascals in Rouen, and filling them with water or red wine in order to measure the height of the fluid balanced by the weight of the air. Seventeenth-century physics was no narrow specialized technical study; pursued in the light of important historical perspectives, free from institutional habits and customs, it offered chiefly a challenge to the ingenuity and persistence of the active amateur. As yet it led to no career, and it brought nothing but the satisfaction of knowing something new for which it was often difficult to envisage any particular use. An exacting discipline, it was advanced with the help of collaborators in far-off places, in Florence and Danzig, in England and the United Provinces, and thus it demanded the knowledge of languages, at least Latin, and very soon Italian, German, English. As different minds turned towards it, there grew up a clearer sense of a boundless frontier to be explored, and of the dependence of the individual worker on the resources and good will of an ever enlarging circle of like-minded friends. In each scientist, in his different frame of reference, in his own language, the sense of fact was developed, along with a capacity for accurate and objective observation, for precise measurement and discrimination, orderly thinking, and increasingly elaborate mathematics. Pumps, lamps, the accounting desk, the clavichord, the musket, the millwheel, all the tools and equipment of the community came to form the storehouse from which the amateur drew his nascent science. To discover a principle that seemed to govern the behavior of these common materials and objects was his delight, and as these principles became more and more consistent, extending towards one another to become laws of nature, and weaving back by inventions into the habits of the trades, the fascination of the scientific quest could only grow.

Emphasis on this creative aspect of science in seventeenth-century France should not cause us to forget that in many ways it was still a largely undefined and rather amateurish enterprise, carried on in an atmosphere of classical tradition, in the light of many ancient and unrefined ideas, with the aid of a multitude of persons whose capacity for science in any exact sense was rudimentary. Although it was a "Century of Genius" in Whitehead's phrase, it was also an age much given to looking back through Renaissance eyes towards the Golden Age of antiquity. The great men were still Archimedes, Hippocrates, Euclid, Galen, Ptolemy, and especially, for many, Aristotle. Even though they were still regarded as artisans rather than members of a profession, herbalists, pharmacists, and surgeons looked up to, and tried to imitate, those who read and followed the classics. New philosophers aligned themselves in accordance with the schools catalogued by Diogenes Laertius; more typical than Descartes' new method was Gassendi's massive effort to write the new knowledge of the age into the framework of Epicurus. In fact, much of the effort of the scientists of the time was directed towards continuation of lines of investigation and speculation revealed or suggested by philological study of classical texts.

The sense of a common effort served in many ways to unite the intellectual elements of the day. The active scientist might be able to explain natural phenomena better than the ancient writer; he could not as yet neglect the basic theories of his predecessor. As he studied and evaluated the classics, scientist or humanist showed certain qualities which came to be taken for granted, but which have become rarer in our day. In his local context, with his immediate preoccupations, he felt that he belonged to a long intellectual tradition which required that he be responsible to mankind, that his science should justify itself through utility and the improvement of man's lot. There was no wish to separate science from technology, indeed the distinction would hardly have been understood. Claude Perrault worked on the text of Vitruvius, because the study of an ancient architect aids, and is aided by, the building of modern palaces. His contemporary, the astronomer and physicist Adrien Auzout, not content with a critical reading of Perrault's commentary, turns a little later to Frontinus *On Aqueducts* to seek methods of restoring the waterworks of ancient Rome for modern needs.

At this distance it is hard for us to draw a clear line between science and humanism, or between technology and the fine arts, as these concepts were understood in the seventeenth century. The principle of the telescope, revealed almost by accident in a glassworker's shop in the Netherlands, was developed and used for science by Galileo in Italy, and had immediate results on the practical level as well as for the theologian and humanist in every country in Europe. The common pumps used in the arsenal at Venice led Galileo and his school to the concept of atmospheric pressure and the invention of the barometer, touching innumerable lines of thought and areas of speculation about the nature of things on the way. Conversely, the builder of waterworks for Montpellier, inspired by a location and a sense of fitness, erected a gem of classical architecture to house his pumps and valves and to symbolize the relationship of the aqueduct of Peyrou to the people of the city.

Perhaps the best illustration of the opposing forces at work in this age of synthesis is offered by Blaise Pascal. Educated in a milieu which cultivated the arts and humanism in a legal background, Pascal evolved full circle from an amateur with a delight in gadgets typical of many in his age to the fully equipped savant who understands what he has found, its theory as well as its practical value and defects. We cannot class all his writings as literature, in spite of the elegant precision of his mature style; most of his works are incomplete, left as their author turned to take up some other more urgent task. But in each he sees his own work clearly, all round it, its philosophical implications, where it leads. His originality stems from this same capacity to recognize and organize fresh material as a basis for reflection and further speculation.

The freedom and flow of his ideas prevented him from becoming a narrow specialist. Far from gregarious as a man, his thought yet turned from the isolated tasks a solitary thinker could resolve, to spread over a continually enlarging area of interest, absorbing and generalizing as it grew. He went far beyond the mathematical basis of his early studies, but the method and the style, once acquired, are not easily lost, and to the end the geometrical manner of expression and proof recurs in all his work. Whether he was discussing effective communication, evaluating statements of fact or theory, or presenting his most daring imaginative flights, as in the *Pensée* on Man's Disproportion, or in the argument of the Wager on the existence of God and the immortality of the soul, the forms and criteria of science are never far away, the traces of positive proof usually present.

Such instances as these illustrate vividly the remark made by Alfred North Whitehead that the modern passionate interest in detailed facts was united in that age with an equal devotion to abstract generalization, to produce what has become an essentially new co-ordinating element in the world. Habitual reference from the external world in which human beings are busy with their daily affairs to a progressively more elaborate system of general principles produced an outlook which the speech of the day described as philosophical, and which may be perceived in most of the literature usually thought of as imaginative. One need hardly stress Pierre Corneille's sudden and dramatic realization in the fruitful 1630's that a play may not only delight but can also seize the conscience of King Public by offering a tragic problem taken directly from contemporary life. *Le Cid* (1636), for all its medieval Spanish elements, derives from its own day, and comments on the claims of present needs and traditional loyalties, on the judiciary and executive aspects of government. Corneille discovers in verse, in the aloofness of the tragic stage, an opportunity for discussion and debate leading to urgently needed decisions in the light of high principles and moral technicalities precisely defined. Corneille could no more limit himself to being a pure dramatist, a mere artisan for the stage, than Pascal could stay within the boundaries of the physicist or the mathematician. Like Galileo or Marin Mersenne, he reflects the new spirit, the new outlook on the world, the new sense of what his contemporaries expect. Aware of current facts and issues in the light of enduring principles, he must explain the moral conflict in his plays in terms that are convincing and consistent, and not merely conventional platitudes. In this application of the intelligence to realms where habit, imitation, emulation of the ancients, routine repetitions, and the tricks of a trade had formerly sufficed, we find something which marks the arrival of a new age in France, and, as a result of France's influence on the taste of Europe, in Western civilization generally. From the sociologist's point of view, or from that of the economic historian, it is innovation: as the seventeenth century recedes into history, it may be suggested that there is something to be gained by seeing an advance that for long has been regarded as purely literary and strictly French in the light of developments in Europe as a whole.

Another aspect of this new outlook on the general problems of the day may be found in the content, the tissue and substance, of the tragic play itself. Having lost the desire to imitate a form, to write conventional rhetoric about ready-made characters in second-hand situations, the dramatists now seek to create an organically new work from the germ of an idea derived from the world of men, from history or accepted legend, nourished until it has developed far enough to satisfy the stylistic requirements of the age. These requirements run deep: the action on the stage must demand conscious decisions of real problems; there must be real people reaching the point of action by deciding between real alternatives. The story can no longer be told in the historical vein of the epic, Aeneas did thus or so, suffering such and such consequences. We must be convinced, as we sit before the stage, that the outcome is still unsettled, that it will be made by the present decisions of the persons before us, that in fact the future is taking form before our eyes. Aware of process in the world around us, knowing that events have immediate causes and inevitable results, we seek in what we read, in science or in literature, a commentary on the processes of life and nature, a rendering of the world we know.

These new requirements on the level of aesthetic satisfaction not only explain the familiar classic unities, they also point towards certain criteria for truth generally accepted in this age. Description must be continuous, the action single, in a given place and a limited time. Once attention is directed towards a given event in human or natural history, the new standard of coherence demands that the observer be informed of precisely what took place at any and every moment of time and at every point of space in the whole region under our attention. In practice, of course, we are content with an approximation to complete information; we remain aware of the possibility that new data may always upset old formulations, and we know that anything present in the space-time framework may be relevant. Erwin Schroedinger remarks that the requirements of complete information and continuity can no longer be satisfied in modern physics, and we may agree that they have no great place in modern art. But this ideal pattern of uninterrupted description, the observational continuum, as a mode of seventeenth-century perception, present in the dramatists and in the natural philosophers, Atomists and Cartesians alike, is typical of the classic manner, subject only to the reservation that selection and arrangement within the space-time framework are needed to meet the limitations of human understanding. It may be suggested that here is an indication of why the unities are important beyond the exigencies of a literary mode. The postulates of science and the conventions of literature coincide to clarify a general requirement of the intelligence.

Still another characteristic of the literature of this age is to be found in its center of interest, the range and limits of its subject-matter. It has often been noticed that French classical literature is, in a special sense, humanistic, that is to say it is centered on man, his interests and activities, to the exclusion of serious attention to the larger cosmic processes as well as to the finer detail of natural history: politics, civil law, the family, the nation, the church; never the universe or the animal kingdom, not even the human race in its many forms. There seems to have been a common decision in France, influenced by the urbane delights of society, by the humane scepticism of Montaigne, by a kind of Pascalian horror at the unknown and unknowable vistas opened by the new scientific instruments, to set a limit on the scope of what could be discussed in polite society, in the theater, and in literature generally. A new astronomy, a renewed physics, a nascent science of living things, the definition of a realm of thought in which mathematics is queen of the sciences, all these tended to reduce the area of imaginative literature to the dimensions of moral man, to his conduct as a distinctively human being, in society, before the law, in his family, as a unit in civil life. The proper study of mankind, as Pope would say, is man; he does not imply that the sciences of nature are an improper study. Rather he suggests by his "proper" that for the generality of mankind the most necessary and appropriate study is the nature and character of man, who is central to the scheme of things, whose interests and welfare are its highest concern, and through whose eyes the universe is known. This is at once a limitation and an ideal for literature, typical of the rationalist, sceptical seventeenth century. It is a rejection of the unlimited enthusiasms of the Renaissance, the boundless interests of a Rabelais, and a deposition against the manifold extravagances of the popular novels of the day, pastoral, heroic, burlesque, and the voyages to the Moon or to Terra Australis Incognita, where wonders beyond those of More's Utopia or Bacon's New Atlantis could be found. It is even a rejection of an important part of the heritage of ancient classicism, of the spirit of Lucretius and the tradition of Ovid, not to mention Plato, Pliny, and Lucian.

These two requirements—that a work of serious imaginative literature be presented with due attention to the criteria of logical proof, and that attention be limited rather strictly to the area within which the moral argument is valid, as opposed to that in which the new scientists were discovering new laws—are each of them products of the new mentality of the seventeenth century, itself the outcome in large part of the scientific revolution. The result was a real division of labor between the scientist and the literary man, not based on mutual repulsion, but the product of mutual understanding, a recognition of the limits and methods of the different outlooks which allowed progress and free movement for each in their respective fields. The total subject-matter of knowledge and speculative thought fell naturally into two great areas; the natural world, from animals through insects and plant life to minerals and planets, subject to natural law, could be discussed by scientists without reference to final causes or the moral law, while the world of man, in which science may be possible but is neither necessary nor absolute, where the irreducible feature is the freedom of the will, demands rather imagination, the *esprit de finesse* as Pascal describes it, intuitive and creative rendering in artistic form, subject to the moral judgment, beyond the realms of precise measurement and accurate enumeration. This dichotomy was not new, but it was reinforced by the peculiar intellectual circumstances of the seventeenth century. And in spite of persistent efforts to bring science into poetry, and the findings of genetics, psychology, and physiology into the novel and the drama, this dualism remains an indelible trace throughout the French educational and cultural heritage.

If the literary mind found itself bounded in certain areas by the kind of knowledge and understanding peculiar to science, it is not to be assumed that this was necessarily a source or symptom of weakness in literature itself. Concentration of attention on a small and limited field meant greater achievement than could have been produced by diffusion of energy over a larger range of themes. Curiously enough, the best literature of France at this time shows many of the qualities of good scientific work; it is complete in structure and relevant detail, in intensity and concentration of interest, in the avoidance of loose ends and gaps in the argument. The drama in particular is characterized by economy of causation, a careful balancing of results and motive forces, an equilibrium which enhances and intensifies suspense. In the prose writing of the day, in the Maxims of La Rochefoucauld, the novels of Mme de La Fayette, there is an objectivity and clarity, an elegance in expression, a parsimony in the argument, resembling very notably certain aspects of science, particularly the mathematical and physical disciplines of the period. How this relationship is to be defined—whether a change in outlook produced both phenomena simultaneously, or whether the vogue of the scientific method produced the reforms which literature needed in all its *genres*—these are questions to be solved only by more elaborate documentation in the sociology of French culture than is now available.

However this may be, the effect of classic texts and taste was reinforced by the enthusiasms with which the mood and patterns of ancient science were taken up by modern thinkers. The advance from the mid-sixteenth century to the era of Louis XIV was marked by a development of sobriety and caution in speculation, evidenced by the number of *Arts de bien penser* in French as well as in Latin published in those years, and by a severe restriction of literary and decorative flourishes, recommended in numerous *Arts poétiques* and *Arts de bien écrire.* It was generally felt that there was a great need to concentrate strenuously on the essential patterns of the material under study; that rational methods, persistently applied, would produce results that neither tradition nor revelation could assure. The chief tendency of the age could be summed up by saying that in drama, as in poetry, painting, the novel, and sculpture, the story was allowed to tell itself by its own internal development, by the elaboration of its essence rather than by the accumulation of external details. Although the art is theatrical, even heavily so at times, yet it possesses a dynamism, and the central psychological or philosophical pattern is fully evoked; *vraisemblance* is sought within perspective and design. Poussin knew how to paint solids on a two-dimensional surface; it was more than a trick of technique with him, it was an application of mathematical theory in the representation of a world which was more than mere appearances. He knew that rational criteria had to be satisfied if his work was not to be merely an illusional device by which the simple-minded would be deceived. In the same way, the dramatist learned that human actions bear a relationship to the momentum of the ages in which they occur. Aware that his audience was as sceptical and as sophisticated as any a dramatist has had to face, knowing that his public was educated on the universalities of Greek and Latin literature, and that while costume and local habits may change, the anatomy of man and the qualities of nature do not, Molière and his contemporaries respected the intelligence and the taste of their day. They might isolate the moral problems, and direct the full vigor of their minds to the discussion of issues of importance in the society of the time, but they would not be forced to exclude consideration of the context, the natural background of process and law in which man lived. Racine presents the death of Iphigenia in terms that satisfy not only the Greek legend that divine intervention was involved but also the modern view—in the mouth of the wily Ulysses—that this miracle only appeared as such to some of the spectators. Similarly, the supernatural elements in his presentation of the prophet Jehoiada may be interpreted in two ways: some can take his great inspiration scene as evidence that Jehovah intervenes to save the Temple against the wicked Athaliah, while others, including Voltaire, can read the text with full confidence that it implies only that Joad acts with consummate skill to create faith in a wavering crowd. The question is not what Racine believed; it is rather how far an audience of the late seventeenth century would accept direct intervention of the supernatural in history and legend. In another vein, Molière could count on the amusement of his public by Monsieur Jourdain's burlesque of the phonetics of Cordemoy, and on their recognition of the comic aspects of the quarrels over antimony and the circulation of the blood. In his work there is little of the venom that characterizes Shadwell's *Virtuoso*, written for a public in which other factors work against the kind of intellectual harmony achieved in France.

The age was one in which the literary artist did not seek out his individual variable. The critics tried to recognize the *je ne sais quoi* of each author, which the author tried sedulously to conceal. Although everyone knew the private gift was there, as peculiar to each as his gait, his bearing, or the idiosyncrasies of his handwriting, the conventions were against singularity, as Molière's Misanthrope discovered. When La Rochefoucauld said that "L'honnête homme ne se pique de rien," he gave a precise and comprehensive statement of the ideal of the intelligent man in the face of his own endowments and activities, applicable as well to the scientist contemplating his discoveries as to the gentleman in peruke and high heels. Happy is the age that can look at the intelligence and the imagination, fact and fancy, and make a comment that goes to the center of the problem of how man shall bear himself in a world which is rapidly losing its human scale.

III

Kin though science and the imaginative arts may be in this age, the branches of the family diverge as the generations come and go, until finally the unity of knowledge has been broken. The founding of the Royal Society of London (chartered in 1662) and of the Académie Royale des Sciences (1666) crystallized the feeling of the scientists that they occupied a special position in the intellectual life of Christendom. At least nominally free of traditional intellectual or religious dogma, they were distinguished by possession of a unique interest and technique, and a particular duty to the community of men of intelligence and philosophical outlook everywhere. Their work was something they had themselves largely planned, they defined their own areas of interest, and elaborated their own methods, to satisfy criteria of their own devising. While a dream of a universal science reuniting a divided Western world still had its fascination for some, the actual course of events in different countries produced different patterns of activity, and science grew in the framework of the political and cultural disunity of the nations of Europe.

In England, the Royal Society soon became large and inclusive, and as a result was only in small part actively scientific. A group of brilliantly qualified workers, Boyle, Hooke, Wren, Lower, Martin Lister, Halley, Ray, Newton, and others, was surrounded by an amorphous gathering of amateurs and hangers-on whose support for science was neither regular nor enlightened. The Society suffered in prestige through having too much said about it in the wrong way. It had an unseemly struggle with the Royal College of Physicians who feared invasion of the medical field by the newcomers, and it suffered at the satiric hands of Samuel (*Hudibras*) Butler, and Thomas Shadwell, who set the stage for later and sharper lampoons by "Sir" John Hill and Jonathan Swift. But the virtuosi did sponsor important work, and their corporate existence fostered the publication of first-rate texts, including notably the foundation stone of the astronomy of two centuries, Newton's *Principia.* A positive influence on literature stemming from science of the Baconian and Royal Society tradition does not arise at once; one of its first products in this kind is the group of poets who use the new insight into visual sensation provided by the *Optics* to produce poetry studied with much insight by Marjorie Hope Nicolson.

In France, on the other hand, the direction taken was quite different, and for a variety of reasons. In the first place, police considerations caused public academic activities to fall under the hand of the generally respected machinery of state. Members of such bodies were salaried; authorities found a more or less suitable place for their sessions, which were kept under more or less discreet surveillance. In the case of the Académie des Sciences, a new observatory was planned, and building operations begun. What lampooning of the new foundation was done was cautious and not at all pungent. The most striking piece of work in this line was the famous *Arrêt Burlesque* written by Boileau and Bernier in support of the new sciences against the decrees of the Sorbonne which attempted to ban all philosophies except that of Aristotle. The opposition of the older men of the Paris medical faculty was sharp enough, but never very well organized nor vigorously pursued. They had long been in a state of war with the Montpellier faculty, with the surgeons, and recently with converts to Harvey's new view of the circulation of the blood. A new royal foundation with not more than three or four medical men in it, even if it was occasionally occupied with human anatomies, could hardly offer much of a threat to a powerful body already endowed with ancient privileges, among them the monopoly of medical teaching in the whole of the Ile de France.

In contrast with the numerous publications stemming from the Royal Society, the fairly regular and rather uneven *Philosophical Transactions*, such works as Hooke's *Micrographia*, and various publications by Boyle, John Evelyn, Ray, Willughby, Grew, Malpighi, Halley, and Newton, the Paris group published slowly and with a caution which outsiders often criticized. A number of pamphlets were published for the attention of the academy, some of them with a false claim to academic approval. But on the whole, the Paris scientists preserved a reticent attitude for a considerable time, and only gradually developed the organs through which publication of scientific work by its members was achieved. The polemics to which Sprat's *History of the Royal Society* (1667) led were thus avoided in Paris, science did not become a matter of public debate, and the Académie des Sciences was able to enjoy uninterrupted prestige and increasing security down to the Revolution of 1789.

While the eighteenth century was to be a critical age in the history of France and of Western civilization in general, it began as if stability had been reached and nothing new was expected to happen. Bernard le Bouvier de Fontenelle had just become perpetual secretary of the Académie des Sciences, replacing J. B. Du Hamel; Fontenelle had published some minor literary pieces and a series of elegant *Entretiens sur la pluralité des mondes* (1686) which had aided in making the system of Copernicus comprehensible to the layman. He began at once to do for the science of France what had not so far been done for that of any nation, not even by the assiduous Oldenburg for the Royal Society of which he had been secretary from 1661 to 1677. This was the preparation of short biographical accounts of deceased members of the Académie, *Eloges*, which, if used with cautious and imaginative discrimination, are still useful to the historian of science and even to the historian of culture. In his way, he was one of the first to see the literary and historical value that could be added to the conventional academic obituary. By including discussions of theory as well as descriptions of scientific investigations and discoveries, Fontenelle made a notable contribution to the expansion of the literary *genres* typical of the eighteenth century.

Fontenelle marks the end of the seventeenth-century synthesis of science and the imagination, and he heralds a new era. His biographical approach to human beings bears a closer relationship to the characteristic patterns of eighteenth-century thought and writing than it does to the dramatic isolation of human problems typical of the age of Racine. While tragedy continues to be produced, its concentrated vigor has left it; one finds the discursive rambling of the Voltairean form, which toys with local color and exotic types in order to illustrate an epoch and a culture (for example, *Zaïre, Alzire, L'Orphelin de la Chine*) while expounding a conventional *philosophe* ethics (*Le Fanatisme*). The real center of imaginative interest is to be found in the novel, whose romantic outlook and bourgeois themes so clearly contaminate the classicism of these unsatisfactory plays. Under the strong influence of Defoe and Richardson, this old *genre* discovers a new vitality, exploits a score of new forms—utopias, imaginary travels, Oriental tales, pretended memoirs—develops rapidly through the whole century, and reflects as it goes all the multiple interests of this sprawling age.

The present point of view allows us perhaps to find a focus for much of this seemingly meaningless dispersion of effort. This is the period in which the French humanist made his first major effort to reform his activities with the methods and criteria of science clearly in mind. The working compromise of the seventeenth century was based rather on the absence of visible divergencies in outlook than on a conscious desire to use literature for scientific purposes or to adapt literary form to scientific content. Now, however, as the long period of relative stability associated with Louis XIV drew to a melancholy close, it was apparent that the scientists were finding subject-matters that were fresh, insights that made a new appeal to the imagination, quite different from the straightforward investigations that stemmed from Archimedes and Euclid. Newton's work on optics, universal gravitation, and the calculus; Borelli and Malpighi on the mechanical aspects of physiology, Swammerdam and Leeuwenhoek on microbiology, and much else: all this was pushing the intellectual horizons of the young man of 1700 further back, and giving him a sense of philosophical complexities quite unknown to the contemporaries of Molière. The study of the natural world was to become a career in itself in a way that no seventeenth-century moralist could imagine; the presence of this outlook, its challenge to the mind, the vigorous impulse and response that they produced in the average eighteenth-century writer make it necessary to look at the literature of the age in a very much larger perspective than suffices for its predecessors.

As we have seen, the classical tradition had contributed a point of view and a mass of subject-matter to literature. Human conflict had been accepted as the central area for development; the subject-matter was history and legend, Greek and Roman, then the tales of civilizations and cultures in contact with Greece and Rome or stemming from antiquity, and finally the history of the church, including lives of the saints. This material was all historically true, or at least probable, and not very seriously questioned; it was useful in that it allowed a speculative mind freedom to discuss human motives, to probe conduct, to consider thoughtfully the main aspects of what the age deemed universal institutions, the family, the state, the monarchy and council, the church, the laws governing conduct and restricting liberty, the customs and manners consecrated by usage. Large as these areas were, they had left little room for the world of nature, for the human being whose interests were not moral, social, or political, but rather directed towards the material universe, as scientist, explorer, practical biologist, navigator, engineer, merchant, manufacturer, farmer, or huntsman. To Molière's Alceste or Tartuffe, even to Othello or Hamlet, Robinson Crusoe was the challenge of a new age.

The transformation of literature that was called for would be almost complete, but the habits of writers and readers change slowly, and reforms were achieved by means of many almost imperceptible progressions. It may be remarked in passing that in France there seem to have been few factors delaying the acceptance of science among the cultivated classes; the Franch satirist does not find the scientist a figure of ridicule, nor does he venture attacks on the royal establishments in Paris or on the innumerable smaller groups pursuing similar tasks in the academies of the provinces. A typical book of the period, marking very well the transition from the seventeenth century to the next, was *Télémaque*, a biographical novel in poetic prose by Fénelon, Archbishop of Cambrai, and tutor to the heir to the throne, Louis' grandson, the Duc de Bourgogne. This popular romance relates the travels and the education of the son of Ulysses under the guidance of the wise Mentor; it has little science in it, of course, but much on political theory and practice, and a good deal about the necessity of developing objectivity in dealing with men and affairs. The essential point for us is the focus on education, on the development of the mind and the creation of useful attitudes in the young man, rather than on an exposition of permanent truths. Fénelon's emphasis is on the unfolding of the mind in contact with older minds, particularly that of Mentor (whose name has become a useful word in English), as well as with a considerable variety of races and organized communities.

Fénelon was no scientist, but he was aware of the trends of his day, and his disciples found it possible to go well beyond his work while keeping the sanity and comprehensiveness of his viewpoint. Chief among these was the Scot Andrew Ramsay, a leader in the establishment of the Masons of the Scottish Rite in France, and one of the earliest advocates of the encyclopedic ideal. Many of his ideas are set forth in his novel, *La Nouvelle Cyropédie ou les Voyages de Cyrus* (1727), which derives from *Télémaque* as well as from Xenophon, although it is much less poetic and idealistic in tone. As benefits a young hero of eighteenth-century romance, Cyrus receives comprehensive instruction in natural philosophy. For the French reader, still under the influence of vortices and the plenum, the physics is Cartesian and pastorally poetic; in the English translation of the same period, Cyrus is given a simplified system of universal gravitation, quite in accord with the prejudices of the insular reader. In one edition of this work, in which French and English texts face each other, a footnote on either page calls attention to the divergencies. Science may be science, but readers are consumers, and their prejudices must not be disturbed.

The two major authors of this first half of the eighteenth century, Montesquieu and Voltaire, both show signs of preoccupation with the new philosophy, and for each of them the impulse is a product of mixed French and British influences. Montesquieu's early project for an elaborate encyclopedia of the sciences is well known; although he quickly abandoned any intention of confining himself to a scientific career, the mark of science is never far from the surface of his work. He may be described as one of those who made a major contribution to the science of politics, and in this way to the sciences of man, to which each of his books, including even the seemingly frivolous *Lettres persanes,* offers something. It is not necessary here to do more than indicate his importance in transferring to the study of political society the methods of rational analysis and causal pattern developing in the sciences of nature.

The evolution of Voltaire's ideas was rather different, but the influence of science was no less profound. Montesquieu had grown up in an erudite rural background, in a family able to assure him of a seat in the Parlement de Bordeaux and a place in the academy of that city. In contrast, Voltaire's background, though legal, was much more modest, and he was much more dependent on his native wit. He could not adopt a new line of thought with any assurance that it might succeed for him, and he could not therefore escape from the cult of mediocre poetry and flat prose to the imaginative vision of a Newtonian universe until rather late in life. Like Montesquieu, he went to England, but as an exiled poet; he was impressed by Newton's funeral, and much more by Swift and Bolingbroke, who were hardly friendly to the scientists of the day. He went home to a life of hiding in France, and discovered that in order to complete his projected book on England he would have to include sections on the sciences, on Newton, the Royal Society, Bacon, and inoculation. Much of this he had to get up for the purpose, and we find him going to Maupertuis for lessons in gravitation and optics. The scientific phase did not last long, but it produced a substantial *Elements de la philosophie de Mr Neuton* in 1738, and it reinforced an interest in the development of the young mind to meet the diversity and impersonality of the external world.

We do not know exactly when *Micromégas* (published 1752) was written, but its contents and temper fit very well with the interest and associates that we know Voltaire had about 1740. It may be debated whether this short novel is rightly classed as science fiction; it is certain that it could not have been written without a good deal of sympathy for the speculation of the day, and in the full light of the extended discussions about the shape of the earth which had rocked, sometimes vigorously, the Académie des Sciences. The giant visitor from Sirius, the less gigantic secretary of the Academy of Saturn, the various astronomers and mathematicians on the ship bringing Maupertuis' Lapland expedition home from measuring a degree on the meridian of Tornea, each represents a recognizable attitude towards the science and philosophy of the day, in a discussion which touches rapidly and lightly on many current intellectual problems. The novel reflects the confidence of a mathematician in the accuracy and validity of his measurements, and his rejection of metaphysics, final causes, subjective impressions, and the like. Micromégas may be derived from Gulliver in Lilliput, but there is a confidence in reason and a scepticism about other ways to belief that is very different from the temper of Jonathan Swift.

The strength of the sceptical position comes out once more in *Zadig*, an Oriental tale tossed off in a few days in 1746. It is a trifle, illustrating the view that men are doomed to disappointment unless they allow for the fact that nature and the external world in general are inherently inhuman in their complete disregard for the interests and preferences of the individual. Like much eighteenth-century literature, this romance describes the education of its hero, ending when he has achieved the mature, rather disillusioned, scientific attitude towards life, already visible in Micromégas, and again later in Candide. A pervasive curiosity, an objective tolerance that finds all shades of opinion interesting and respectable as long as they do not interfere with liberty of inquiry and belief, a systematic pursuit of truth in spite of traditions and doubts—these, much more than a taste for sentimental botany and rhapsodical astronomy, were the product of the five years or so that Voltaire spent in active pursuit of science at Cirey with Madame du Châtelet; like his heroes, he has learned from science, and achieved in his own way a synthesis, quite different from that of the seventeenth century.

It is not necessary here to review all of Voltaire's work. In him, as in Montesquieu, and in Denis Diderot and Jean-Jacques Rousseau, we meet a mind whose perspectives lie almost wholly within the framework of the natural sciences. Each of these reaches a view of man and society by an inductive process from credible facts, assuming that a natural origin for law is possible, even probable. Civilizations are described in such a way as to lead men to sounder and more useful decisions on the basis of facts available to all. This utilitarianism marks a significant advance over the typical products of seventeenth-century thought. It exerts a powerful and often unacknowledged influence over much that is to follow, in the writing of history, in the social sciences, and in the imaginative forms of literature and criticism.

In this way the contact between literature and the sciences grows progressively closer as the classical and theological patterns of human nature lose their validity for the reading public and the writer. Old forms will no longer hold the new data, the motivations of Racine please but do not convince a reader of La Mettrie and even Buffon. A new didactic literature delights and instructs a generation of readers who need no romantic plot to enliven a picture of rugged nature in its untamed aspects. In another vein, we have the series of dialogues in which Diderot recounts D'Alembert's dream, informing us of the latest theories of generation. In England, Erasmus Darwin, and in France, a better poet, André Chénier, develop a new style in poetry, using the materials and ideas of science as a means of satisfying a new generation of readers. J.-S. Bailly, the historian of astronomy, and Condorcet, the theorist of the idea of progress, prepare the way for the arid Abbé Delille and the brilliant Shelley; after several years and much groping, many of these tendencies are worked into a structure of another kind, Victor Hugo's *Légende des siècles* (1859–1883), in which episodes of human history chosen for their moral and dramatic value in the long ascent from the Garden of Eden to the *Great Eastern* and the dirigible balloon are presented as stages in the conquest of the world by the growing intelligence of man.

With the nineteenth century this fusion of science in the literary imagination is accelerated. We note the scientific ambitions avowed by Balzac in the Avant-propos of his *Comédie humaine*, the impassive study of the varieties of human nature given us by a Mérimée or a Stendhal, the influence of Alexander von Humboldt on two generations of poets who come to see nature through the eyes of the observational biologist. Lamarck teaches Sainte-Beuve, and thus contributes to the foundations of modern literary criticism by suggesting a method of analysis and classification of authors according to the *famille d'esprits* to which they belong. A new pessimism in poetry develops as the place and destiny of man on the planet and in the cosmos are seen in geological and astronomical perspectives; this is reinforced as the definition of what man is, his biological nature, is determined by the tradition that runs from Cabanis and Bichat through Magendie and Broussais to Claude Bernard, and thence into the literary hands of Emile Zola. In the long run, literature is invaded, even pre-empted, by men who accept science because they can find no tolerable alternative interpretation of the facts of life. Against this situation, the conservatives, clergy, layman, rich and poor, rage in vain; and down to about 1900 the futility of their reaction to what they describe as scientism, secularism, or positivism is demonstrated by the absence of anything resembling a masterpiece that is not inspired and informed by, or at least consistent with, the trends arising from science. When Baudelaire, in *Les Fleurs du Mal* (1857), revolts against much of contemporary materialism, he cannot escape the emphasis which an expanding science of psychology places on the experience of the individual, nor avoid the criteria of coherence and logical consistency towards which all thought still tended. His language, for all its symbolism and synesthesia, is predominantly concrete and particular, capable of precise understanding in its own frame of reference. The conclusion of the book, with its defiance of death and *ennui*, tends to enlarge the area of rational discourse rather than to eradicate or emasculate it; the final quest is "Au fond de l'inconnu pour trouver du *nouveau*!"

Aware of the long history of the earth and the short chronicle of man, of the infinite expanses of space among the stars, and the fragile nature of life in a hostile universe where densities and temperatures range in inconceivable scales, the nineteenth-century writer could no longer utilize or even respect the socially centered superficialities of the classical tradition. Old myths are parodied—*Orphée aux Enfers*, *La Belle Hélène*—or they are retold in a new archaic vein by poets like Leconte de Lisle, Mallarmé, Paul Valéry. Both phenomena are evidence of a new aesthetic freedom deriving from the wider perspectives of a new age; the style of classicism has lost its magic as men have become aware of the infinite number of possible patterns in which an old story may be told, the tremendous variety of ways in which its human values may be brought out. The value of a legend is no longer in the legend itself; it gains power when man uses it in a new way to test his own capacities of creation and evaluation. The librettoes of Meilhac and Halévy, the operettas of Offenbach, are hardly matter for an inquiry of the present order, but the Parnassian poets are, and their understanding of the truth of art is a subject to challenge the analytically minded critic. History on the basis of scientifically criticized artifacts is the end-product—and the terminus—of traditional humanism.

IV

Begun by G. B. Vico, developed by Herder, applied by Mme de Staël, elaborated by Sainte-Beuve and Taine, the Romantic revolution in taste and appreciation was at the outset an appeal against the mechanical view that man's activities were caused by the play of simple basic instincts which any moralist could codify. It led, as time went on, to a new science, a "Scienza Nuova" as Vico had declared it would, and it sought allies in the systematic studies of man. On one hand it was literature, music, all the arts, poetic, creative, dramatic; on the other it was a renewed world of learning, opening up fields of philology, anthropology, sociology, new areas of history. The revolution affected every aspect of life, from religion and politics to the use of leisure time, and its monuments are all around us, in gardens, churches, the plans of towns, and the furnishings of our houses. It inspired speculation about origins and destiny, and it set a generation of young men on the path to a new knowledge of history and the condition of man.

Once freed from the classical notion that the business of art was the preservation of ancient forms, the commemoration of legend and history, literature developed in rhythm with the enquiring spirit of the age, radically different from the antiquarian temper of the professors of rhetoric, even in the Sorbonne. The poet could turn his attention to experience directly, without the need of seeking its expression by means of traditional models and vocabulary. He could approximate ever more closely to the known forms in which the cycle of birth and growth, reproduction and decay are cast; he could convey a sense of the universality of nature in the particular occurrence, with no preconceived idea of what that universality might involve. It is no wonder that the typical novel of this new tradition, *Gobseck*, *Madame Bovary*, *Germinal*, is a case history, the novelist a biologist, the typical hero more clearly perceived as an animal than as an immortal soul. The slant, once acquired, is not easily escaped; the idealists in reaction, Paul Bourget, Maurice Barrès, use naturalist techniques and justify a conservative and humanist political and philosophical position on biological grounds. They too stem from Taine and Claude Bernard, from Lamarck, Humboldt, and Vicq d'Azyr.

The trend, as the nineteenth century drew to its end, was towards the analysis of consciousness into its components, the exploration of lower levels of perception and feeling, the discovery of what lay beyond the normal threshold of awareness in dream and trance. Into this psychological pattern fit many of the experiments of the symbolists, trying to do with words what contemporary painters were doing with colors, modern musicians with sounds. Rather than up towards an ideal, the onlooker is asked to look more closely at the buzzing, blooming confusion of the nascent mind. Here the emotions and a glimmering intelligence reach towards form, and artistic creation begins. Mallarmé's Faune comes to life, and writes scraps of a Sicilian idyll; he is not so much a symbol as a complete and fully imagined organism, living on a level rather below the evasive dream that teases the imagination of the reader. He is much more satisfactory to a biological age than the Centaur of Maurice de Guérin or the Satyr of Victor Hugo, antecedents if not models of Mallarmé's creation.

Much of the best writing of this period embodies a protest against the taste of the day, against predominant trends in popular literature, against tendencies habitual in the thought and sentiments and prejudices of the increasingly important reading public. Symbolist poetry is as much as anything a manifestation of this protest; in an atmosphere of beauty the Faune represents—but how differently!—the same preoccupations and interests as Zola's peasants. He reflects equally an interest in the lower, less articulate kinds of intellectual power, enhanced psychologically because he is capable of formal discourse in his *poème*, presented as it is in a background of dawning consciousness. *Bateau Ivre*, best known of the poems of the boy genius Arthur Rimbaud, is another example of the same type of reaction; a derelict canal barge is a thing of small beauty in itself, but treated subjectively, from within, by a single vigorous imaginative effort, and elaborated by means of associations, it becomes a vehicle for a vision of the natural world, enhanced by the very violence of the contrast between commonplace appearance and sordid function on the one hand, and the color and grandeur of tropic skies and landscapes on the other. In each case the poet extrapolates from data actually present by way of his reading; both Mallarmé and Rimbaud understood the necessity of rooting a myth in an aspect of experience where objective knowledge could lend authenticity to the structure of the vision. For Mallarmé, the Faune derived from legend and early Greek literature, even though his eye may have been stimulated by a painting by François Boucher; there are echoes of Theocritus and Bion, a glimpse of the landscape of Sicily, the psychology of the primitive, the unfiltered sensations of the human being delighting in sunlight among marsh plants round an unsullied pool. Rimbaud draws deeply on his schooling and his culture; the *Nautilus* of Jules Verne, the *Monitor* of the American Civil War, the sea life studied in the marine biological laboratory at

Roscoff, echoes of a dozen books, of readings and woodcuts in the *Magasin pittoresque*, all combined to fill his stanzas with the world as the nineteenth century could see it when it cared to look. Dream world and fantasy though it all may be, it has its roots and substance in the serious knowledge, the science and technology of the age.

A full explanation of either poem would be very much longer than the poem itself. In each case, the critical reader has never failed to be impressed by the strength of the internal fabric of these poems, the absence of wilful injections of personal fancy and random deviations from the logical and dramatic structure. The dream pattern is there in both; but it is the Faune or the Bateau Ivre who do the dreaming, not the poet, who stands aside from each creation, impersonal magician of his *sorcellerie évocatoire.*

The background of this symbolistic poetry becomes apparent as we move away from it towards new conceptions of art and literature. We can now see that movement *L'Art pour l'Art* did not die when Parnassianism was replaced by the vigorous ferment of Symbolism; and we can suppose with considerable confidence that the cult of artistry retained certain features of the outlook of the scientist, his objectivity, his interest in the precise recording of appearances, his insistence on the potential relevance of every detail. Parallel with changes in the direction of scientific and intellectual interests generally, the arts—poetry and painting in particular—changed the direction of their outlook; the poet turned now towards a more careful examination of his instrument and its capacities, the use and qualities of his language, and explored the relationship of language to the world of his perceptions.

The nineteenth century was an age in which the intellectual and the artist tended to retreat behind the barrier of his specialty. Finding little to please him in the bourgeois industrialism of his time, with all its brutal ugliness and smug complacency, the poet withdrew to a world in which the intelligence, the artist's as well as the scientist's, could offer a coherent vision of reality whose relevance was not denied by experience. Traditionally, Symbolists and Parnassians offer one of those convenient pedagogical antitheses beloved by the systematic historian of any literature; but as a matter of fact we can see now that just as those two old polarities Voltaire and Rousseau have more in common than once was commonly taught, so these nineteenth-century schools of poetry rise in a common background of ideas and sentiments, and differ only in the means they choose for expression of their perceptions. Both groups distrust classical aesthetics, the imitative academic vein in art, quite as much as they react against ineffective romantic emotionalism and its appeal to vulgar and uncritical prejudice. Both grow on lines parallel with contemporary intellectual interests, the new scientific study of languages, archeology, psychology, sociology, the history of trades, sciences, and the arts; each of these areas belongs historically to the humanities, was once part of the field studied by the Renaissance humanist, and was now being abandoned to the new specialists with scientific method. These poets of the late nineteenth century in different ways, some protesting, some making an effort to reclaim lost ground, try as they can to maintain the unity of their universe by means of the poetic intuition, the comprehensive insight, which is the chief instrument of creative art. It is a mistake to deny them at least an intellectual purpose, a program of action in the light of the new outlook on man and his world visible in contemporary philosophy and science. A deep distrust of old methods, an abandonment of older ways of feeling and believing, an anxious quest for new means of expression and communication, along with a vigorous debate over where new and better techniques could be found, invaded all fields and produced, is indeed still producing, the variegated chaos we have to take for granted today as we think of modern movements in art, and the new fields opening for the younger scientist.

The position of Baudelaire in the earlier stages of this development is unique and generally regarded as of the utmost significance. His secret, so far as it can be penetrated, seems to lie in his imaginative use of all the perceptions at once, in a fusion and moulding of data from every level of consciousness. Awareness of bodily needs, hunger, heat, cold, fatigue, pain, and of raw sensation seems never to have overwhelmed his philosophic sense of the wholeness of man. Perceptions from all levels are co-ordinated in most of his poems, discrimination between each permitting him to see and communicate a symbolic relationship, a *correspondance* as he calls it, transcending all the data.

In this way he is far from being a poet of mere sensationalisms. More than his predecessors, and often much more than most of his followers, Baudelaire draws with precise insight on the whole range of human consciousness as the material for his poems. Through his hunger or his passions he may feel the suffering of others, but he proceeds further, beyond sympathy, to generalize. He does not linger as does an Alfred de Musset to glorify the individual reaction, the private delight that sympathy indulges. Suffering, and with it the revolt against suffering, are characteristics of the human situation; Baudelaire, like the contemporary naturalistic novelists, finds the individual case representative of a recognizable order of things, not permanently beyond all human reason nor ultimately mysterious. Man lives in a large world, full of *correspondances* which may be clarified in the long run by awareness and intellectual effort:

> La Nature est un temple où de vivants piliers
> Laissent parfois sortir de confuses paroles;
> L'homme y passe à travers des forêts de symboles
> Qui l'observent avec des regards familiers.
>
> Comme de longs échos qui de loin se confondent
> Dans une ténébreuse et profonde unité,
> Vaste comme la nuit et comme la clarté,
> Les parfums, les couleurs et les sons se répondent.

There seems little reason to believe, on the basis of such poems as this, and there are many, that Baudelaire differs from other authors of his day in being either obscurantist or a mystic. Fullness, richness of perception, consciousness of the difficulty human intelligence faces as it tries to make sense of the increasing complexity of the data presented by the refinements of modern techniques, is not the same as religious or mystical contemplation that denies differences in the light of a single unifying principle. The All may be seen by the One; but for Baudelaire, the All is not One—if it were, then why should there be *correspondances*?—nor is the One All.

Baudelaire thus marks the beginning of a movement away from optimism, either of the positivist scientific variety or of the sentimental romantic stripe. Yet he has not lost faith in man; as poet, he believes in the arts, even if they are lighthouses of despair; in his philosophical meditation on human destiny, he holds out hope in the great journey of the race, not in the achievement of a final goal, but in its value as experience, as a means of showing what man can do, of testing man's own powers. The introspection he suggests is humbler and less deceived than that of the great Romantics, Chateaubriand and Hugo; it is also more acute, more critical, better controlled by standards of evidence and coherence. It has validity even when it rejects naïve common sense as offering criteria for the judgment of the self and its laws. Baudelaire proposes an intuitive approach to the problem of truth and reality which is not accepted or understood at once, but which has its deep roots in the thought of his time, and which, with Bergson and William James, will become part of the heritage which the nineteenth century passes to the twentieth. No scientist himself, Baudelaire's work is generated by an age of science, of faith in science; his is one of the most intelligent reactions to the development of contemporary thought, and no account of the relations of science to literature in France, or indeed in Europe as a whole, would be complete without it. It is such work as his that raises the level of the discussion above that of mere carping and recrimination; he brings up the basic issues, of value and meaning in the full human context, of human dignity and self-respect, in such a way that the need for a solution is imposed, a solution that others also will seek and approximate, if they do not actually find it.

There is perhaps less need to discuss the scientific background of French naturalistic fiction, chiefly because the relationship is well known and very simple. Novelists of this school, mostly followers of Emile Zola, adopted the techniques of natural history, the careful observation of individuals and groups, extending the method as the human material required so as to include not only the instincts, passions, physical and physiological conditions responsible for much human behavior, but also traditions, sentiments, loyalties, and ideals, whether praiseworthy or not. Such a descriptive technique implies a serious effort to classify the various factors that govern human conduct. Etienne Lantier, hero of Zola's *Germinal*, is shown systematically on several different levels of interest; the plot is not significant in comparison with the picture of a human being, his drunkenness, his hunger, his control over the mob, his love for Catherine, his social ideals, and so forth. In Joris-Karl Huysmans' complex novel, *A Rebours*, another hero, unheroic Des Esseintes, with equally conscious purpose, is composed of traits selected from every level of human character; his thirst, his unscrupulousness, his acute sense of pain, his vices minor and major are presented in the same scale and structure with his endless pursuit of an elusive, aesthetic ideal. The emphasis in each case is on the human being as a whole, and not on the resolution of a subtle moral problem, as it is in classical literature. Huysmans remarked, in a preface written twenty years after *A Rebours*, that the great service rendered by naturalism had been to situate real persons in exact surroundings; he added that this vivifying influence had made it impossible, in theory at least, to avoid the average, which in the long run made for dullness and mediocrity as the technique came to be imitated.

The ultimate model of these writers was the scientific monograph. Raw material, the evidence of the author's own experience, carefully supplemented by the use of scientific documentation, treated with a minimum of subjective interference, in series of events corresponding as closely as possible to the known sequences recorded in the daily newspapers, with a general view of working out the probable results of combining selected forces operating in ordinary life—this was what was meant by the phrase "experimental novel." There is thus, in theory, no relationship between the personality of the author and the meaning of his book, as the Romantics had held. The sense that there is an inevitable individuality of style in thought itself, in the choice and arrangement of materials, in method, seems rarely to have occurred to these authors; for them a physicist is not involved in his work, and regularity of result, the formulation of law, bears no relationship to morality or even temperament. In the eyes of the naturalist, the sole object of interest is the book itself, the report on the experiment, which leads us through various stages of analysis and objective interpretation to a clearer understanding of the nature of man, and the sequences in which his affairs are seen.

The break with classical tradition is now complete. The character-types described by Theophrastus, still useful to Molière, had been compared to animal species by Balzac, whose vivid imagination saw foxes and wolves, lions and sheep, cows and parrots, donkeys and serpents, even more clearly than La Fontaine, in the common run of human beings, in the tinkers and money-lenders, notaries and chemists, farmers and medical students on the highways and streets of France. On this simple and observed analogy Balzac erected a completely erroneous genetic pattern for his treatment of the family; the naturalist novelist soon saw that species in the scientist's sense of the word do not exist among men, that family patterns obey other laws, and that a causal chain, if it is to be found in this area, must be based on other data. It was not discovered at once that a human being is in many ways unique; it was however soon seen that the combination of factors operating in any one person could be extraordinarily complex. A man may be forced into common patterns by society, but the drama in each life lies in the struggle that goes on endlessly between the individual and the forces that act on him from without. Zola's hero, Etienne Lantier, is a miner, a leader, an intellectual; he is hungry, weary, jealous; when enough statements have been made about him, he becomes unique, and much of the naturalist theory is overthrown.

The human race, then, as the naturalist finally has to see it, is one, without real cleavages between men, with no barriers that cannot be broken down, once conditions are favorable. Contrariwise, unfavorable conditions can set man at war with man, brother with brother. The hero of *La Débâcle* kills his dearest friend in the Commune; the hero of *L'Œuvre*, absorbed inhumanly in his painting, destroys his wife and child, and sets himself apart from his community. And in another naturalist, Guy de Maupassant, tragedy follows from the excessive development of the vices, avarice, jealousy, vanity—traits normal and useful enough at a low level among human beings, but disastrous unchecked.

Thus the naturalistic novel illustrates the application of biological principles to life, not only in the sporadic use of medical and genetic principles, but in the general approach to the problems of plotting and the arrangement of the causal chain that underlies each story. The insistence on *vraisemblance*, on honesty in describing men and societies, on impersonality in style, all are factors in which the scientific influence is unmistakeable. The central figure or group is studied fully and objectively, on the basis of original observation, in itself, in relations with others, in the habitat, and in the normal activities of life and milieu. The idea is old, it had occurred in Diderot's desire to see plays of this type; and it satisfies an interest which has become characteristic of our society, which determines taste in most of our popular literature, in the movies, television, and the radio.

V

In an article in his newspaper, *Combat*, November, 1948, Albert Camus sums up the modern age by saying that the seventeenth century was the age of mathematics, the eighteenth that of the physical sciences, the nineteenth that of biology, and "Our twentieth century is the century of fear." Fear, he adds, is not a science, but science has had a hand in the matter, since its latest theories have led science to contradict itself, and its practical advances menace the earth with destruction. Furthermore, even if fear is not a science, there is no doubt that in unscrupulous hands fear is a technique.

The epigram does not suggest that science has disappeared from the scene or from consciousness; on the contrary, it suggests that in defense of its own sanity, the twentieth century has had to begin to live with science, alarming though the prospect may be. In literature, both aspects of the process have been described, and the tensions and anxieties are perceptible. Balzac, Flaubert, and Zola claimed scientific validity for their work, asserting that in part at least their books could be regarded as an annex of science. "Science" was a good word for them, as for their age; "scientific" expressed approval, and pointed towards the realization of desired ends.

With the passage of time, the words "science" and "scientist" (the latter suggested in 1840 by the Rev. William Whewell of Cambridge University) have lost some of their savor. Although twentieth-century man no longer sees a promise of inevitable happiness in science and its works, he cannot deny science its importance, or refuse the necessity of understanding it. In literature, practically no school or considerable author today regards science as offering a necessary or even a particularly important analytical or expository method, or an ideal or valid criterion for aesthetic or moral judgment. In certain *genres*, in dealing with certain aspects of human affairs, the technique of using scientific language is one of many skills to be acquired and adopted, valuable in analyzing certain sequences and patterns of events. Thus at various times scientific words have been useful; one recalls the vogue about 1900 of the idea of evolution and Ferdinand Brunetière's application of the word, misleading and erroneous as it was, to the history of literary *genres*. Other words and concepts, from psychology and medicine (*complex, fixation, symptom, diagnosis*) and from physics and chemistry (*electric, lens, magnetic, catalyst, filter, synthetic, crystallize*), have sometimes been useful in permitting a more exact analysis, a new look at the sequences, the processes and struggles, of the kind of material an author uses, and have undoubtedly encouraged a degree of impersonal objectivity in so far as such phrases have been able to replace old and well-worn—and therefore vague—formulae.

In other respects, science has permitted, as Professor Deutsch has suggested, the exploration in literature of new areas of human interest, and thus has become a province of the literary heritage, subject, like all other parts of that increasing storehouse, to the ebb and flow of popular taste. Plays like that of François de Curel, *La Nouvelle Idole* (1899), whose title suggests its ideological slant, and Bernard Shaw's *The Doctor's Dilemma*, brought scientific discussion in more or less dilute form to the modern stage; the novel by Paul Bourget, *Le Disciple* (1889), uses a highly melodramatic plot as a basis for a full, if somewhat biased, discussion of the morality of modern science.

The chief phenomenon, however, in this area, is Jules Verne, whose long list of scientific romances sprang from a serious effort to anticipate the technological applications of the sciences currently in full development. Submarines, airships of several patterns, countless mechanical devices, are suggested and utilized in his novels, in such profusion that these books in turn are known to have influenced the scientific vocations of many young men, among them Simon Lake and Georges Claude, and the navigators Charcot and Bernard Franck. As do most of his contemporaries, Verne writes in positivist terms; the fantastic element is only apparent, a more advanced engineering, using the principles and tools of his own day rather than the marvellous and irrational inventions of a Cyrano de Bergerac. But neither the society nor the technology anticipated by his creative writing corresponds to the world as we know it today; for all its inventive fertility, its technological ferment, Verne's mind could not foresee such an innovation as the internal combustion motor, or imagine such problems as those imposed on the aviator by weather and atmospheric conditions. Twentieth-century science is stranger than nineteenth-century fiction; Verne remains well within the bounds of scientific possibilities, and no longer surprises his readers with his extrapolations.

Paradoxically enough, it is this very restraint that has prevented Verne from becoming recognized as a serious literary figure. Reticence may be a scientific virtue, it is not always a literary one, and a generation that delighted in highly seasoned literature relegated Verne's books to the popular and schoolboy level, with disastrous results as far as his style and general literary quality were concerned. Science penetrates literature by becoming literary, less pedantically technical, more attentive to the broader implications, and less insistent on the scrupulosities of laboratory techniques. Biological determinism in the Zola manner ends by being dull; the hypothesis precludes effective dramatic struggle, and the full use of creative imagination. The fine balance the effective novelist must maintain between an increasing understanding of the conditions that govern the phenomena of living, and the part played, *per contra,* by unpredictable individual impulse, is essentially the source of dramatic conflict, and the human basis of all interest in literature. From this point of view, science, as used by most conventional authors, even by the professed naturalists, is merely an extension of "common sense," the general knowledge, or in some cases the mental and moral philosophy, of an earlier generation.

In this respect, there is much resistance in the twentieth century to science as an absolute frame of reference, as a source of all the answers. Life as the novelist sees it is mostly non-scientific, not to say pre-scientific. Nor is there any prospect of an immediate reformation of the literary or humanist part of humanity; for them, as for Pascal, the word "natural" condemns the biological and physical side of humanity as being rather less than fully human. Out of this stems the *angoisse,* the tensions or anxiety, of much modern writing in France.

Although set in modern terms, the conflict is a very old one. Whether scientists or not, most French thinkers are deeply attached not only to the idea, and the ideal, of humanity, but also to its realization, present and possible. Thus the Frenchman fears any radical change that may threaten this idea, whether it involves a loss of accepted human qualities, a break in the tradition, even in the tradition of free expression, or radical adventures within the concept itself. The French author—and Verne is typical in this respect—hesitates to speculate about powers unknown to man as he is at present; he rarely introduces a character who possesses capacities in any way superhuman, and then only in such a way that he may escape the accusation of being ridiculous. In each case where a strange and non-human being is presented—Micromégas, the Centaur, the Satyr, the Faune, the barge in *Bateau Ivre*—the pattern of the mind, allowing for necessary extravagances of number and dimension, is human; the relation of sense and will, the relative importance attached to the different senses, the direction of intent and purpose, as found in humanity generally, is never questioned. In many respects, this attitude resembles the naïve universalism of classical science, based on an assumption that all phenomena can in the long run be explained in a single set of terms, that the mind need never be at a loss or make a gratuitous assumption in achieving knowledge. But this outlook, philosophical or "psychologique" as a

Frenchman would call it, has occasionally severe limitations, especially in the acquisition of general understanding of peoples and ideologies which do not belong to the Greco-Christian tradition; there seems, for instance, to have been a good deal of distortion of Oriental philosophy before it was generally received in French literature. Even the understanding of the thought of such near neighbors as England and Germany suffered considerable delay; but understanding of philosophy long preceded comprehension of both Shakespeare and Milton for neither French taste nor French patterns and measurements were quite ready for such heady draughts. And as science advances, towards an Einsteinian universe, for instance, resistance develops more acutely in France than elsewhere.

Much might be said in terms of this essay about the literature of the first half of the twentieth century. Positivism does not lose ground, but it finds new competition as it becomes more and more a conservative and academic orthodoxy. Anatole France believed in nothing very much, but he retained a special kind of sceptical respect for the sciences. Georges Duhamel, doctor of medicine with a significant record of actual practice, introduces not only scientists, but also scientific concepts and issues into his novels; in *La Nuit d'orage* (1928), in the cycle of the Pasquier novels (1933–44), and other works, he makes a conscious and important effort to strike a balance between the claims of science and the intuitions of the humanist as a means of directing conduct and evaluating the course of events. Jules Romains also owes a good deal to his training in biology; in *Les Créateurs*, volume XII of *Les Hommes de bonne volonté*, he describes with remarkable insight and sympathy the process of scientific discovery, in perhaps the most strikingly original and penetrating passage of this sort in any modern literature. It should be noted in passing, that in none of the works of these authors is the conflict in values and interpretation placed on a religious basis; always the issue is between science and what is perhaps to be regarded as a larger kind of science, the general intuitive life of a widely read humanist.

Closest to the general tradition of modern science is the work of Antoine de Saint Exupéry, aviator by profession, but a well-qualified mathematician and physicist and a writer of considerable ability. Apart from his novels about commercial and wartime aviation and two fantasy parables (*Le Petit Prince* and *Citadelle*), *Terre des Hommes*, whose text and title were both inadequately rendered into English as *Wind, Sand and Stars*, exhibits a close contact with science. The message of the book is a protest against war and a eulogy of humanity in man. With a background of geology, astronomy, and meteorology, Saint Exupéry presents the human struggle for knowledge and security against the powers of the natural world; with an anthropologist's understanding of tools and techniques, he compares the airplane and the plough, the flyer and the shepherd, and evaluates the long process by which the tool, even when it is as complicated as an airplane, comes to be taken for granted and used with the simple directness of an Indian's paddle. Creation for Saint Exupéry, whether of a perfect airfoil or the B minor Mass, is all one; it is the chief end of man, and a civilization that stamps out human beings in series and deprives millions of individuals of an opportunity of creative growth is a hideous mistake. This fruitful and comprehending book is perhaps a key to all his work; it is at any rate one of the sharpest indictments of an era in which science was allowed most unscientifically to destroy humanity in large areas of the civilized world, and with it science itself.

Saint Exupéry speaks for an age, for men like André Gide and Malraux, for Paul Valéry, and countless others. His goal, like theirs, was to establish an idea of man, to solve the most important problem in contemporary thought, to discuss what man is in the light of present knowledge of the universe, its magnitude, composition, structure, duration, and changing states. The urgency behind this question has been imposed by science more than by any other branch of human activity; and the need of discovering a sound and generally acceptable answer, in the light of all the data, was never so pressing. Saint Exupéry, like most of these men, would agree that the basis for an answer must involve not only communication among all men but also understanding and agreement, and the subordination of attitudes and loyalties that impede agreement.

There are thus, at the end of this long development, signs of a trend towards balance, towards the achievement of understanding as between science and the various obstacles it has met. Western society has found that it cannot long endure without science; and Western scientists have found that they cannot go far without certain features of society as the West has understood it. Albert Camus, for one, in his novel *La Peste* (1947) has illustrated most of the tensions and anxieties which have plagued these two centuries of development. He shows the limits within which science can operate freely and with assurance in the complex impersonality of the modern world; he presents knowledge, not as an absolute, but as symbolic, a construction of the mind rather than as something given. Man is complicated, in constant need of understanding his own wholeness; his problems are difficult largely because of the extreme difficulty he finds in setting them clearly, of avoiding their own indeterminacy. The issue is restated in *La Chute* (1956), and the conclusion is not essentially different. What man can do is a matter of his own choice; as he becomes free, as he recognizes the dignity, the worth and values, implicit in his own situation, his destiny will not fail him.

As Camus, and before him Saint Exupéry, could see, in an emergency the ends are simplified and emphasized, and technical method becomes an accepted and undiscussed means. Science, as free inquiry, ceases; technical perfection, a useful art, is subordinated to an end which is in the mind and conscience of everyone. Anxiety, an essential ingredient in literature, in the dramatic struggle of epic, theater, and fiction, accompanies life at its highest tensions; in this mood, the calm impersonality, the objectivity of the scientific temper are less useful than the skills, the techniques science offers for life's struggles. If science grows by asking How? rather than Why? then at these moments the imperative nature of the Why?—the reason for action—thrusts the How? into the background. Progress therefore is not automatic; only when the question of ends—the Why?—can be submerged, accepted unemotionally and removed from consciousness, perhaps by the ending of the state of anxiety, can the vigilance and discipline that science requires be turned back to the advancement of disinterested learning. The dichotomy of literature and science is thus invincible and inevitable. Anxiety and tension are the parents of literature; objectivity, impersonality, a lack of ethical concern, a certain hardness in the face of the facts, are the breeding ground of science. Only rare and infrequent junctures in human history, such as those we have found in the seventeenth century in France, and perhaps again in modern times, permit fruitful collaboration on a common task.

It may be, then, that if this chronicle contains any hidden message, it is a challenge to each of us to promote conditions in which our precarious control over ourselves as human beings and over the impersonal forces of nature allows us to meet anxieties and tensions with clarity of vision and ethical understanding.

IV

The Creativity of Science

by
DAVID HAWKINS

THE DEVELOPMENT OF SCIENCE; LANGUAGES, WAYS OF COUNTING, ABSTRACTION, QUANTITY; FROM TELEOLOGY, TO MECHANISM, TO EVOLUTION; LAW AND WILL; MAN'S IMAGE; THE PARTICIPANT OBSERVER; THE GROWTH AND AUTONOMY OF SCIENCE

I. SCIENCE IN THE MODERN WORLD: ITS HISTORY

In the last few of man's thousands of generations science has made possible a new society; new in organization, size, and technology, new in medicine, and new in its arts of war. In historical perspective, says the historian Herbert Butterfield, our world "is stranger than Nineveh and Tyre." It is his judgment that no innovation in history is comparable to that of modern science in either the mass or the variety of its consequences.[1] Our world is strange, half familiar and half alien, even to us who inhabit it, as probably to their inhabitants Nineveh and Tyre were not. In our concern to understand this world and the revolution which has led to it, we tend to identify science with its technological antecedents and consequences. This is an over-simplification which may be allowable for some purposes, but it conceals the nature of science as an expression of human creativeness, and thereby it conceals the most important humanistic aspects of science.

The evidence of our failure to understand science is found primarily in the modern humanities; in literature and the arts, in history, and in philosophy. The history of science, to begin with, is a relatively neglected subject. Consequently, but more importantly, the general historians treat science not at all or in detachment from the rest of history. The philosophy of science is not so neglected, it has an old tradition. Indeed the whole of philosophy since the Renaissance shows a deep influence from science. But philosophy in general, and the philosophy of science in particular, has become technical and needs as much interpretation as science itself, if not more. Literature and the arts have given a vast amount of attention to the machine and its promoter, and to all the liberating and degrading aspects of industrial society. Still there has been a neglect of science. The technician and the engineer and the research worker, the very people who live intimately with science, have been inadequately recreated within the world of the arts. The scientist remains largely a stranger to those who would learn about him in the perspectives of novel or drama. Yet without some such portrayal the scientist cannot be understood adequately; even, very probably, by himself.[2]

Neither the machine nor the Industrial Revolution are aspects of science. These things imply the existence of science, and they imply an alliance between science and industry. This alliance has been elaborated, moreover, until it is now a complex working association. It is certainly no mere external association of convenience, for it springs from the character of the allies. Science in the last hundred years has developed new aspects because of its alliance with industry, and whether they are "humanistic aspects" or not, they are of the greatest human importance. But science has not become the creature of industrial interests, it has its own interests and its own way of doing things. These interests and ways of doing things have to be understood first, and the place to discuss the symbiosis of science with industry is towards the end of the account, not at the beginning.

Long before the industrial age the intellectual innovations flowing from science were a source of both jubilation and mistrust. One may go back from the nineteenth century to the seventeenth, and find there a different sort of impact of science on society and culture. But the influence has continued and spread, its course marked by such landmarks as the world-machine cosmology of the age of Descartes and Newton, the atomistic materialism of the eighteenth century, and the biological materialism of the nineteenth. The story of influences and interactions would offer a fairly complete account of the development of modern culture.

Yet these influences and interactions are just that: in themselves they are not humanistic aspects of science. As is the case with industry, the connections of science with liberal culture have been rich and intimate, and to understand science one has to explore these connections. Indeed, the telling and retelling of the story of science and the modern world is an important discipline of the modern mind, and a student who does not know much of this story has not much of an education. But still, one does not get an adequate account of science from the usual point of view labelled "modern cultural history."

On the whole, the role of science in the usual story of modern culture is rather stilted. Science is there in every scene, performing some momentous feat of discovery. It is treated very respectfully, and sometimes with awe and reverence or at least with the kind of enthusiasm that may pass for awe and reverence. There are some critics who do not share this enthusiasm; they excite a good deal of sympathy sometimes, but of course they are quite wrong. Actually the critics of science have this important virtue, that they make it come alive. The official eulogies, on the other hand, accord science the treatment usually reserved for famous admirals and ex-presidents in the movies. It is better to be presented as a villain, in some part of your true character, or even to be misrepresented as a lively and believable rascal, than to appear as a plaster saint or a wooden Indian or a man from Mars in a space-helmet.

What is lacking in the official accounts is a sense of the inwardness of science, of its creative impulses and its secret ambitions, of its irrationalities and its intellectual growth, of its pettiness and its greatness—in short, of the human character of science. If the creed of the humanist indeed be expressed by the maxim, *Nihil humanum a me alienum puto*, then the essential humanistic aspects of science are to be found, not in the method of science if there be such a thing, and not in the results of science as they appear in textbooks, and not in the external influences of science on industry or on politics or poetry or painting, but in the life of science as an expression of human capacities and limitations. In saying this, one claims for science only the same protection against distortion and irrelevancy that the literary humanist claims for poetry or novels or dramas: that they are to be understood first of all as works of creative perception, rather than in terms of their social effect or the personal idiosyncrasies of their creators or anything else.

II. SCIENCE AND THE HUMANITIES

In the first of his Terry Lectures of 1946, President Conant of Harvard suggested that science is most clearly distinguished from other intellectual activities by its cumulative character, its capacity to make a kind of indisputable progress. The suggestion is certainly not a new one; in the battle of the "ancients" against the "moderns" to which Mr. Priestley alludes in his essay, this progressive character of science was the main argument of the "moderns." What distinguishes Mr. Conant's treatment is the subtlety and informality of his description of the strategy and tactics of science, by which it has progressed. His accounts of various case histories of scientific progress are believable, which is not always true of formal abstract statements of the scientific method. Mr. Conant thinks that the distinction between the sciences and other intellectual activities is rather sharp. He refers to "subjects as diverse as mathematics, physics, chemistry, biology, anthropology, philology, and archaeology" as having made indubitable progress in the last three centuries. He continues:

> A similar statement cannot be made about philosophy, poetry, and the fine arts. If you are inclined to doubt this and raise the question of how progress even in academic matters can be defined, I would respond by asking you to perform an imaginary operation. Bring back the great figures of the past who were identified with the subjects in question. Ask them to view the present scene and answer whether or not in their opinion there had been an advance. No one can doubt how Galileo, Newton, Harvey, or the pioneers in anthropology and archaeology would respond. It is far otherwise with Michelangelo, Rembrandt, Dante, Milton, or Keats. It would be otherwise with Thomas Aquinas, Spinoza, Locke, or Kant. We might argue all day whether or not the particular artist or poet or philosopher would feel the present state of art or poetry or philosophy to be an advance or a retrogression from the days when he himself was a creative spirit. There would be no unanimity among us; and more significant still, no agreement between the majority view which might prevail and that which would have prevailed fifty years ago.[3]

One may suspect that Mr. Conant's evocation of ghosts would bring some surprises. The philosophers, at least, would feel some professional obligation to question the question rather than answer it, and against four such adversaries Mr. Conant had better be well prepared with definitions and elucidations. There are after all several obvious senses in which painting has progressed from the time of Michelangelo, as well as several senses in which the term "progress" is simply not applicable, so that to say painting has not progressed is like saying that the number two is incombustible or that equality has no mass. If cumulativeness, not further defined, is the mark of progress, then surely poetry has progressed; for we have the poems of Dante, Milton, and Keats, and also those of more recent poets. If the question is rather whether Emily Dickinson is a better poet than Donne, then there is a similar, and no clearer, question about Einstein and Newton.

The real point of Mr. Conant's comparison is that he wishes to call attention to certain peculiar characteristics of the scientific tradition. Similarities and differences between science and the arts or philosophy will carry the discussion a little way, but only a little way, and such comparisons are, almost inevitably, going to become invidious. What is important is to consider the cumulative character of science in its own setting.

Definitions of science suffer usually from being too broad or too narrow. If the definition is broad enough to include all the relevant aspects, it includes many intellectual activities that are not ordinarily conceived as scientific; and if it is narrow enough to exclude such non-scientific activities, it excludes part of science as well. Thus a great many writers appear to regard the power of prediction as an essential characteristic of science. On the one hand, however, the power of prediction is as widespread as human life, and on the other hand there are sciences in which it plays no very central part, as for example biological evolution or the theory of numbers.

The difficulty of "defining" science in this sense is not particularly a difficulty to be overcome; it is an expression of the facts of the case. Even at its most primitive, human life is impossible without the degree of science—ordered experience—essential for the use of language, stone tools, and fire. With the appearance of urban society, the organization and extension of such knowledge became more deliberate and more specialized in certain areas, notably mathematics and astronomy. We find the first plentiful evidence of systematic inquiry, freed from domination by immediate interests, among the Greeks; and Greek philosophy, in particular, created the vision of a world open to investigation by disciplined intelligence. The intimate interweaving of theoretical interests with the interest in experimentation is predominantly an achievement of the modern world. Just where, in all this, does science begin? All the elements and their interworkings are there in pre-scientific common sense; what has changed is the intensity, and the degree of specialization, of certain intellectual activities, and the extent to which they receive social recognition. Almost one is tempted to say, that is science which is pursued and recognized as science. Almost, but not quite.

This deliberate and institutionalized use of knowledge to extend knowledge need not imply a rigid, uniform method, a set ritual of inquiry. The methods of inquiry are more stable than its results, but the distinction is one of degree, and new knowledge brings new methods. If the attempt is made to prescribe a single exhaustive scientific method, adequate for all times, the result is likely to be trivial or false. On the other hand the conception of science as cumulative knowledge has certain definite implications as to methods.

To be cumulative, to show conspicuous and persistent growth, a science must be systematic in its formulations, organized for maximum suggestiveness and guidance in seeking new data, and for maximum relevance to their interpretation. But organization of scientific knowledge means more than suggestiveness and relevance, it means to make positive claims beyond the evidence that is at hand, and put these claims at the mercy of fact. In this sense a science must be objectively verifiable. Its formulations must ultimately be tested directly or indirectly, against regularities in nature which are independent of the desires of individuals or social groups.

Verification is essential to a science, but it is by no means the only essential. Since it can never be complete, it serves as a principle of self-correction rather than as evidence of an achieved finality. Science cannot afford to reject a whole formulation because of isolated contradictions between it and experience, until it has achieved a new formulation retaining the essentials of the older one. For this reason single experiments or observations are seldom crucial, although they may appear so in retrospect. In the search for reformulations there is often room for casuistic arts. Verification is essential to growth, but orderly growth may require freedom from the immediate indications of experience. Particular facts are the ultimate criterion of scientific truth, but they do not disorganize uniformities already found.

The discipline of science is thus twofold; the first part is the discipline of systematic coherent formulation, the second is that of empirical verification. These disciplines determine the character of scientific creativity as the discernment of new analogies and their evolution into conceptual patterns which, in turn, bring with them new anticipations of experience. The discipline is different from that of other arts and inquiries, and the creativity differs in like measure from that expressed as poetry or architecture or drama. One may agree with one humanist that there is no useful purpose in spreading the word "humanities" as a label for all creative activity. One may certainly accept the "versus" that so often comes between "humanities" and "science"—provided it be understood as an expression of the variety and the aggressiveness of creative activities, and not as equivalent to a contrast of the creative and the routine, the organic and the mechanical, the humanistic and the merely factual. There are conflicts within the humanities as well as between humanities and science, as for instance between poetic and philosophical interpretations. The fact of such conflicts is no evidence of unrelatedness!

The recognition of the creativity of science places it alongside, if not among, the humanities, facing the great universe that includes man and would dwarf him except for his capacity to question and affirm, to accept the universe as a challenge requiring new insight and new mastery. In this the sciences and the humanities are one, their adepts share equally the creative experience, what Emily Dickinson described as

> The moments of dominion
> That happen on the soul
> And leave it with a discontent
> Too exquisite to tell.

One may agree also that the human essence so conceived does not derive from the conception of man as a consumer, or as a bundle of biological urges or conditioned reflexes. But man has all these attributes, and his imperfect description of himself in terms of them is a creative achievement of the same order, although not of the same species, as his painting or drama or music. It is not man as the subject of these descriptions, but the descriptions themselves, which exhibit the humanity of science. The humanist's perception is truncated and distorted if these creations, being human, remain alien to him, and if he does not see them as they are and for what they are.

Science and art are equal, but not identical, assertions of man's attempted mastery over what would subdue him. The barriers to this view are a number of incomplete analyses of the two domains, which when taken as complete and adequate have the effect of transforming the difference between science and art into an absolute antithesis. One of these, already discussed, is the contrast by subject-matter, emphasizing the "what" rather than the "how" of scientific knowledge, the textbook rather than the inquiry that produced it. An analogous view of painting or music would be that which identified their highest product, in every generation, as the pastiches of the schools. A textbook is a pastiche of science. Textbooks have their place and their own standards of excellence. They are tools of the art, and where they assume undue importance they destroy the scientific spirit. "The letter killeth; but the spirit giveth life." The highest product of science in each generation is the research of that generation, expressing its new departures within the traditional frame, its definitions of new problems or its fresh attacks upon old ones.

A second partial contrast between science and art derives from the observation that the artist's work is an expression of experience, concerned as such with the unique and the concrete. The contrast then is with science as an abstract and discursive treatment of general relationships, which ignores the concrete and the unique. The contrast is not false, but it is misleading. On the one hand, the notion of art as expression—even universal expression—obscures the essential traditionalism of the arts, never seen more clearly than in periods of metamorphosis, when new forms and old standards have to be measured against each other. The contrasting notion of science, on the other hand, is equally misleading, that it merely accumulates in ever neater form the humdrum regularities of nature; this notion obscures the constant preoccupation of the scientist with the particular phenomenon as expressive of a reality that is never wholly docile, never exhausted by the neat descriptive paradigm within which the scientist would contain it.

A third related contrast is that between the diversity and individuality of works of art, as against the cumulative and essentially anonymous character of science. The scientific worker has as much need of withdrawal as anyone; but his private creativeness is under a special group constraint. He must do his work in a way that others can repeat, and he must describe his activities in language essentially prosaic and literal, so that others can test or continue them. A piece of science may be complete in the sense that it is competent and conclusive, but the final test is repeatability. A piece of poetry can hardly be a "piece" in that sense. The poet achieves immortality when his poem, the specific artifact, is preserved through generations. The scientist achieves immortality by having his work paraphrased into the textbooks of later generations, with perhaps a biographical footnote thrown in.

Again the contrast is misleading. If Newton's ghost were asked specifically about the progress of physics he would have, probably, a generous appreciation of nearly three centuries of progress on many fronts. But he might justifiably observe that deep problems implicit in his *Principia* had been unduly neglected for over two and one-half centuries, partly on account of too little reading of the work itself, and too great reliance on the virtuosi who had paraphrased it in elegant and condensed form. Einsteinian relativity more or less takes up where Newtonian relativity left off, and until Einstein very few physicists had understood Newtonian physics as well as Newton did. If "a poem should not mean but be," the opposite assumption about a work of scientific genius can be a very unsafe one.

III. THE EVOLUTION OF EXPERIENCE AND OF STYLE

Science is essentially cumulative and essentially social. Individual creativeness in science would have itself judged, and in the end is judged, by its contribution to the heritage of organized knowledge. The success of science as a whole is measured by the acceleration of the rate of growth of knowledge compared to the slower and more spontaneous changes of its cultural milieu. This contrast in rates of growth is a familiar one, but it must not be made absolute. Science is at once a product and a mode of cultural evolution, and the philosophy of science is from this viewpoint a chapter in the philosophy of human nature and culture. We have alluded to the one-sided view of science that puts it in false contrast with the arts. This view is best corrected by describing science in the setting of cultural evolution.

The advent of Darwinism threw the old debate over the nature of man into a new form. Once it had been impossible to discuss man without using the language of a static dualism; in man, soul and body, angel and animal nature, struggled for supremacy, and every moralist elaborated the fierce ineluctability of the warfare. The *Origin of Species* and the *Descent of Man* were followed by substantial discoveries of archeology and anthropology. These, and a wide rereading of the literature, revealed human history in new outlines. It extended much further into the past than the written record had so far allowed, and through the greater part of its course, it was a story of biological development. But in the later stages, cultural evolution was seen to have been incomparably more rapid than biological change; man's animal nature and his spiritual essence were related, not as antagonists within a single unalterable frame, but rather as the more constant to the more variable, as stable to nascent, as structure to potential.

Biologically, man is an animal apt for culture. For a time, a very long time, the aptitude and its expression paced and limited each other in reciprocal development. At some point, which we need hardly define here, biological limitations ceased to be critical, and cultural changes accelerated sharply. Tools now extended man's bodily powers and dimensions, and language communicated and refined his needs. V. Gordon Childe, in *Man Makes Himself* and other writings, has emphasized the detachable and extra-corporeal nature of tools, which allows them to develop faster than the hand that makes and uses them. Similarly, language, once invented, permits experience to accumulate socially without reference to the time-scale of biological adaptation. In both physical and spiritual dimensions, society becomes the vehicle of whatever in man is distinctively human.[4]

Described in these terms, human nature is not a static thing, but a process, a dynamism. It is not the ability to use tools that distinguishes man from his nearer cousins in the animal kingdom, but an ability to produce the tools he uses. Indeed, it is not until further development springs from this last trait, when tools are used to make tools, that a growing culture begins to free itself from biological restrictions. When late eolithic man began to spall flint seriously, using fire and water or special bone tools, he entered on the first slow stages of a chain-reaction whose later products include the turbo-generator and the atomic bomb. When tool produces tool, perpetually new functions and continually improved performance become possible, and the spiral of civilization is on its way. Artifacts eliminate the rough stone and the piece of driftwood, crude natural objects that suggest their own uses; these now become raw materials, prized not for their uniqueness of form, but for the humbler quality, uniformity of substance.[5]

A similar dynamism develops on the spiritual side. The growth of language leads to the capacity for abstract thinking, thinking without reference to an object physically present to the speaker. It also frees thought from the limitations of individual private experience, and provides a social vehicle of experience. "Language," said Hegel, "is the public consciousness." Abstraction produces the habit of generalization, in which reference to an object suggests reference to classes of similar objects. Through these tools of the mind, in their own way, like physical tools, detachable and extra-corporeal, experience is freed from the restrictions of animal communication and readied for the impact of new perceptions. The solving of problems, a phase of development described most elegantly in the studies of William James and John Dewey, is here part of the essential mechanism of intellectual growth. Generalizations, taken from the lore of the group and modified by individual experience and application, give the individual an enormous extension of power, enabling him to anticipate consequences in new situations and, in turn, to provide for the growth of his knowledge. Primitive man is a creature of action, who does not value knowledge for its own sake; his interest is centered on the practical problems before him. Perhaps. But whether he is aware of it or not, every application of his knowledge tests it and, potentially, extends it. The search for knowledge is latent in the very use of it.

Although culture embodied in the invention and transmission of tools and language is essentially dynamic, it is relatively unchanged over the life-span of individuals. It may appear completely stable and unchanging to those who participate in it. Tools and techniques liberate man through control of the material environment, at the same time imposing constraints upon him. Some of these constraints derive from the physical properties of tools and materials—including the sheer cussedness of things. Others are indirect, stemming from the culture as man finds himself expressing it. Working at a new problem, the artisan is conscious only of the material and the tool, their powers and limitations; within these bounds, he operates freely and attains his results in any way he can devise. But when he faces a recurring situation, his work is disciplined by a sense of style, of established patterns of doing things. Creativeness is not destroyed; its expression is the affirmation of value and viability in the patterns of culture within which it has worked.

The concept of style is often taken too narrowly, as though it were characteristic only of the fine arts, of the way a writer writes or a painter paints. You can hardly persuade a good machinist to do cruder work than his machines and his skill permit; the job may not require a close fit, but good workmanship does. This is a matter of style. In a very similar sense there is style in the layout of parts in a radio chassis or a gasoline refinery, and it is for this reason that all objects of utility may be judged as objects of art. No product of human labor can be purely functional.[6]

Style, as an aspect of human activity suggesting structure, makes any culture seen from the outside appear as relatively rigid and unadaptive. Style is an abstract thing, as we usually speak of it; its social embodiment has the form of an institution, a social expedient devised as a way of meeting a recurrent problem, accepted and taken for granted as a habit. Devised as a means, such a creation is often transformed into an end; the "why" of a social organism does not include all its "what," all its significance in the context. Ways of life are elaborated into special disciplines and loyalties by those who participate in them; the sense of style goes beyond the pragmatic conceptions of efficiency into conduct more related to the maintenance of those disciplines and loyalties than to the specific ends originally served. One has only to contrast the rain dances of the Pueblos of the Southwest to the seeding of clouds with silver iodide in the same area to perceive the varying extent to which pragmatic goals may determine social conduct. Legal tangles around cloud-seeding suggest, moreover, that considerations of a more stylistic order are not wholly absent from the exercise of modern methods, however pragmatic they may be in their immediate intent.

Style and its institutional embodiments are of the essence of culture. They are conservative and stabilizing in their influence, as determining ends, and not simply instrumentalities of living. Indeed, to those who participate fully in it, a culture must appear in its basic aspect as beyond challenge, as rooted in the nature of things. This picture of culture appears to be in sharp contrast with the easy stereotypes of science, as revolutionary, with clear goals which are somehow defined once and for all in its original charter. Yet knowledge, formulated in language and transmitted in the schools, may also be thought of as institutional in character. The abstracting process, describing particular events and qualities in a systematic pattern of concepts, gives private experience a form ready for social currency, and makes any generalization in some degree independent of the individual who gives it expression. Even though a judgment or general idea may be completely empirical in origin, it can easily lose all particular reference to the experience from which it arose. As an element in the knowledge belonging to a culture, it is subject to the variations of opinion and interpretation, restatement and debate, typical of the culture and character of its people. The process of reformulation and extension takes place in the full context of prior generalizations and intellectual attitudes of the culture as a whole; the tradition of thought is adapted or extended to accommodate the novelty, and the institution of knowledge in the culture becomes visible rather as a way of accepting subject-matter than as the intellectual content of the culture itself.

Organized knowledge, of which science is the most typical example, leads a life of its own as a component of culture. Deriving from experience, it is larger than experience; it has a style, and the accompanying loyalty to style, which experience as such does not have. The recognition of this autonomy explains the frequency with which one encounters a doctrine of science for science's sake in certain fields of basic research. In such a field as the theory of numbers there are many whose confession would be that of the monk in *Boris Godunov*, "I write for myself and God." In fact the theory of numbers has no very important applications except to other branches of mathematics; yet it is one of the fields most highly regarded among mathematicians, whose criteria in this case are intellectual-aesthetic, not utilitarian.

The concept of style in relation to science is not radically different from the concept of style in the literary or plastic arts. The example of mathematics offers parallels to the variety of literary *genres* discussed in the essay of F. E. L. Priestley. Among the Greeks there was a certain *genre* of geometrical analysis which related the study of plane and solid figures to ruler-and-compass construction. The restriction to ruler-and-compass construction was one of those stylistic constraints, not imposed by subject-matter, which defines an approach to subject-matter, and even implicitly limits it. Such restrictions are on a par with the style of the *tema con variazioni* or the sonnet. Malraux employs a very similar conception of style when he says,

> Cette signification des styles nous montre, par un puissant grossissement, comment un artiste de génie . . . devient un transformateur de la signification du monde, qu'il conquiert en le réduisant aux formes qu'il choisit ou à celles qu'il invente comme le philosophe le réduit à ses concepts, le physicien à ses lois. Et qu'il conquiert d'abord, non sur le monde même, mais sur une des dernières formes qu'il a prises lui entre les mains humaines.[7]

But we should add to this, or qualify it by the insistence, that nature is not infinitely amenable to the forms which an artist or scientist may choose or invent. Changes of style may open up new areas of subjects, or the concern with new subjects may force the development of new styles. The content of Conrad's *Nostromo* could not be put in epic verse, nor can modern geometry be rendered adequately in the style of Euclid.

Because knowledge, seen from this point of view, leads a life of its own, in part and at least in the short run, independent of empirical discipline, philosophers have long puzzled over the relation of empirical subject-matter and rational form. The age of an institution, its stability and the sheer mass of experience on which it rests, give a participant in it a sense of its eternity, of absence of change. Similarly, the analysis of knowledge reveals axioms, first principles, which appear self-evident, indubitable, beyond the range of doubt or question. The illusion is the same in either case; in a stationary approximation, knowledge must be viewed as deriving from experience only in detail, owing its more general structure and validity to inherent rational principles which in turn owe none of their authority to experience.

It is now widely recognized that such principles are, indeed, independent of experience in a very real sense. Not matters of fact, they concern the style of our knowledge, the manner in which facts are accumulated, discussed, and ordered at a particular time in a particular culture. Immanuel Kant was the first to give a detailed philosophical exposition of this view; he presented it as an outline of the forms of perception and judgment deemed to be inherent in the human mind. This, the "critical philosophy," was presented in the framework of the conceptual patterns as Kant found them; in itself it was still a stationary approximation, accepting principles as fixed and given, prior to experience, and therefore absolute. Later thought has not accepted this restriction, but recognizes the evolutionary character of concepts and categories, changing in the general evolution of culture. Examples from the philosophy of Kant are the conceptions of Euclidean space, Newtonian time, the principle of causation—all of which have been subtly, but radically, transformed in the history of science since Kant's time.

A science—any science, or science as a whole—develops out of common sense knowledge. Its growth involves two assumptions: first, that there is more to know than the context admits; and, second, that there are assured ways by which the unknown can be discovered. This sense of an unknown and this assurance of method are of course already implicit in the background of culture. The important change is one of emphasis, a reversal of ends and means. Primitive men used knowledge as a tool to solve immediate problems, and as a by-product their knowledge was slowly enlarged. In science the immediate problem is a test-case selected as a means toward the deliberate enlargement of knowledge. The recognition of this shift of emphasis, the building of institutions to protect and encourage it and give it prestige, constitute the real beginnings of science. In this way, science is a second major acceleration of human evolution; it is a development compared with which general cultural evolution is leaden-footed, and it is infinitely faster than biological change. From the latter part of the seventeenth century, observers, friends and critics alike, have noticed repeatedly that, as they say, science has advanced more in the last fifty years, or the last century, than in all the previous eras of recorded history.

This rapid growth of science within a cultural envelope, which in turn changes more slowly within the conditions of life and the material environment, has been possible because, in a measure, science has been free to develop in accordance with its own needs and nature without external constraints. The culture itself, as it advances and becomes conscious of itself, develops play in its points of bearing. What the machinist calls its "tolerances" increase; it loses the close dependence on the habitat which is typical of primitive societies. Within it, perhaps the freest of all its components, science advances at its own pace, in its own direction, or directions, and maintains what speed it can. It is not surprising that from time to time the growth of an infant science causes discomfort and painful dislocations in the cultural matrix that supports it, and anxiety and distress to the consciousness of the world around it.

IV. THE DEVELOPMENT OF SCIENCE

One of the great contemporary physicists, Niels Bohr, introducing his discussion of the conceptual aspects of modern physics, briefly describes the most essential features of scientific procedure in these words:

> The task of science is both to extend the range of our experience and to reduce it to order, and this task presents various aspects, inseparably connected with each other. Only by experience itself do we come to recognize those laws which grant us a comprehensive view of the diversity of phenomena. As our knowledge becomes wider, we must always be prepared, therefore, to expect alterations in the points of view best suited for the ordering of our experience. In this connection we must remember, above all, that, as a matter of course, all new experience makes its appearance within the frame of our customary points of view and forms of perception.[8]

The Ordering of Experience: Languages and Ways of Counting

An ancient way of thinking of fire, still dominant in common language, may serve to illustrate the kind of ordering of experience to which Bohr refers, and the way in which it may be revised in the face of new experience. The word "fire" functions as a name of a material, and for a wide range of usage this mode of functioning is adequate. Nevertheless the history of chemistry shows rather clearly that the description of fire as a substance comparable to earth, air, or water inhibited the growth of understanding for a long time, and was only set aside as the result of exacting empirical discipline and long debate. On the other side of the ledger, the history of thermodynamics demonstrates with what caution old ideas must be laid aside in science, and provides by example a needed corrective to the common notion that the history of science is a history of the discovery of error, ending miraculously in the present with what is true. Necessary as it was, the abandoning of the caloric (substance) theory of heat was a step backward in one respect, because the caloric theory had contained the idea of the conservation of the quantity of heat, whereas the kinetic (motion) theory of heat at first lacked any such principle, which we now know as the conservation of energy.

It is always easy to see in retrospect the ways in which habitual patterns of thought have both inhibited and supported the growth of knowledge. This same systematic character, it goes without saying, provides the sole basis for all analogical thinking, and thus for efforts to find recognizable order in new ranges of experience. The inhibition of new knowledge is not a result of systematization *per se*, but of unconscious systematization, resulting in inability to distinguish between the order of nature which experience reveals and the order of ideas which has become habitual in a given culture. The analysis of such connections of ideas and with it the discipline of logic has been one of the essential components of the evolution of science.

The development of elementary mathematics provides the example of the analytical discipline. The first persistent and successful effort to disentangle the systematic connections of ideas, it has stood as a model and inspiration ever since. Despite the analytical character of mathematical truths, referred to above, it is possible to make out a sense in which such systems as arithmetic and geometry may be regarded as empirical sciences. They are not mere records of experience, it is true, but they have evolved along with, and out of, the recording. The basis in experience for these disciplines is, however, of so primitive a character that we can only with difficulty recapture the relevant passages of experience, or describe them without apparent circularity, presupposed as they are by the very structure of our language. Nevertheless ordinary arithmetic is applicable to the description of nature only where, and to the extent that, we deal with relatively discrete and relatively permanent objects. Number ideas are usable only because there exist groups of objects of one sort or another which can be counted, which if counted in different ways give the same "number," and which can be classified as "equal to" and "greater than." If a primitive fisherman makes a knot in a string for each member of his clan, and catches a fish for each knot, then everybody can have one fish. This is one empirical counterpart of Euclid's axiom, that "things that are equal to the same thing are equal to each other." It is a fact of experience, and could not be otherwise. A world without some such facts is not "thinkable"; in such a world, our "facts" of arithmetic, facts of a purely logical order, would not be false but they would be irrelevant.

Geometry likewise rests upon empirical foundations which are, psychologically speaking, self-evident. That is to say that our ordinary experience of spatial relations is already in a sense implicit in the language by which we describe them. Yet it is a fact of experience and not of logic that spatial relations possess–to a high degree of approximation–that Pythagorean property which Euclid assumes, according to which the square of the hypotenuse of a right triangle is equal to the sum of the squares of the other sides.

The growth of urban culture, and with it of more massive architecture, surveying, the weighing and measuring of commodities, and the development of monetary exchange led to a more complex use of arithmetic and geometry. New conceptual expedients and new techniques were devised: angular measurement and triangulation, the plumb-bob, the abacus, and, above all, written notation. "'Formulas" were discovered, reducing the solutions of complex problems to simpler ones whose solutions were easily obtained. The mathematics of the valley civilizations correspond most closely to those of modern engineering handbooks. Modern engineering rests on a foundation of fundamental science; but while it would be foolish to suppose that in Egypt or Mesopotamia there never occurred what we would call scientific curiosity or mathematical insight, it is almost certain that these things gained for themselves no distinctive social recognition. "Pure theory" was no doubt there –some of the Mesopotamian discoveries might argue as much mathematical genius as those of later time—but sociologically it was implicit and spontaneous, not defined or encouraged as a special division of human labor.

Nevertheless the practical knowledge that had been gained in the Nile Valley and Mesopotamia had only to be transferred to a new setting to provide rich material for what we now call theoretical mathematics. It is certain that many of the theorems of geometry attributed by the Greeks to themselves were already known long before.

Abstraction, The Language of Quantity: Limitations of Greek Science

The Greeks discovered that there are strong logical connections existing among the multitude of geometrical facts. Euclid's *Elements* exhibit the extraordinary implications of this discovery, that a small number of elementary and independent geometrical principles clearly stated lead by the newly developed technique of logical deduction to all the rest. It can indeed be argued that something like this type of relationship among geometric facts was already implicit in the engineering practice of the Mesopotamians and Egyptians, who knew in many cases how to reduce more complex to less complex problems—for example, relating volumes in three dimensions to areas in two. But reduction in complexity is only one type of analytical procedure, and not the most fundamental. Perhaps it is safe to say that the distinctive contribution of the Greeks was to recognize that theory could be developed apart from practice. It was they who first diverted the path of science from the field and the shop to the study. Some contemporary writers have tended to reproach the Greeks for "divorcing theory from practice." It is true that they did not steadily appreciate the significance of turning periodically from laboratory to shop and field. But it is hardly right to speak of divorcing theory from practice when the theoretical interest was only just emerging as a well-recognized part of the human enterprise.[9] No doubt many Greek natural scientists and philosophers exaggerated the power of rational analysis. Only thus perhaps could its limitations be discovered. A certain withdrawal from empiricism might well have been needed, to be followed by a return to bolder and more fruitful application. The early Greek scientists succeeded in the withdrawal, but their successors after 200 B.C. increasingly failed to carry out the return, often even to attempt it.

Classical geometry, based upon an engineering tradition of known facts, led in turn to the formulation and solution of many new problems, extending far beyond the range of previous experience. These discoveries led in two directions, toward new applications and toward new abstractions. Both Hellenistic and modern science involved an extension of the range of experience and imagination inconceivable without techniques of quantitative discrimination. The often-expressed ideal of science as quantitative—implied in the invidious label "exact" science—had its origin in these extensions, and its limitations are to be found there too. Unfortunately, a good deal of the ordinary praise of quantitative description in science has lost sight of the origins and limitations, and as a result great amounts of scientifically useless information have been accumulated and decently filed away, merely because the information in question lent itself to description with two-decimal accuracy. This is doubly unfortunate, because the notion that the language of science is inherently quantitative is one of the main sources of confusion about the nature of science, and, mistakenly, one of the main grounds of protest against the scientific mentality. Thus the engineer who looks at the waterfall only to translate its sublimity and power into kilowatts, thus Whitman's learned astronomer, and so on. Thus, on a more philosophical plane, Bergson's rejection of "spatialized" time as told by the clock. A careful study of the way in which quantification has been bound up with the progress of science will throw light on, and will do much to qualify, the belief that the ideal of science is a world of pure number and abstract form.

The concept of quality implies a judgment of likeness or unlikeness which is achieved by some process of matching order and relation in simple objects as well as in extended problems of style. The quality of a phenomenon is recognized by comparing it with another phenomenon, present or remembered, and the style or order of a complex of things is recognized by comparison with other complexes. Quantity is measured by second-order matching. To say an object is green is to say it is in some sense like other green objects. To say there are ten green objects in a collection implies that in scanning it several such judgments were made, and that the objects or judgments as a collection have a further similarity to a set of verbal counters. The recognition of quality is the beginning of all knowledge, and numbers themselves are qualities of collections.

In the history of chemistry qualitative accounts precede and underlie quantitative accounts, just as drawing precedes and underlies the study of proportions in art. Each statement about quantity modifies some preceding recognition of quality, and is otherwise unintelligible. As long as science continues to grow there is always a range of problems and processes where qualitative recognition is possible, but where quantitative discrimination is not yet achieved.

Above the level of simple qualities and uniform measures the relations of quantity and quality become more intimate and complex. When we speak thus about style, order, and relation, the simple contrast is misleading. Consider as an example the cone of ancient geometry. Regarded in one way it is a simple order or pattern, a *Gestalt* that can be grasped as directly as any quality. Yet implicit in it is the whole complex of quantitative features of ellipse or hyperbola that the ancient geometers sought to disentangle. For all that they "murdered to dissect" the *Gestalt* in question, it comes alive again, and indeed with new richness of association, as soon as analytical attention is shifted away from its details. The poetic validity of the image of the earth's shadow-cone, in the poem of Tennyson to which F. E. L. Priestley refers, depends upon the preceding history of quantitative discriminations in geometry and astronomy.

And this really is the point, that quantity and quality are poles between which knowledge in its growth must alternate. Quantitative or late-stage knowledge leads to qualitative or early-stage knowledge in new fields. Chemistry became quantitative in the time of Lavoisier and Joseph Priestley. That it became so was not an end in itself; it was only the starting-point for the atomic theory of Dalton and the image of molecular architecture developed by Kekulé. We can just as well speak of quantitative knowledge as early-stage, and qualitative knowledge as late-stage. Indeed in the realms of the very great and the very small this reversal is entirely appropriate. For in these realms the grasp of the qualitative features of nature—very different from those of the man-sized realm—is a late achievement, based upon often minute and apparently unimportant differences among the phenomena directly accessible to observation. The inverse square attenuation of gravity, a fact which determines the whole character of the astronomical universe, amounts only to a minute difference within the range of direct terrestrial experience. Kepler's discoveries depended upon the full accuracy of Tycho Brahe's observations of the planets, and Newton's analysis resulting in the law of gravitation was first an inspired guess based upon Kepler's work.[10]

It may appear a stretching of words to speak of Newton's gravitational law as expressing a grasp of *qualitative* features of nature. For is not

$$F = Gmm'/r^2$$

the very epitome of all that is quantitative? Not if one takes seriously the moral of the cone discussed before, which has as complicated an algebraic definition. Like any other language, that of mathematics acquires simplicity and richness, the capacity to represent complex qualities, only through use and association. That Newton's law is an algebraic schedule for numerical manipulations is only the lowest level of its meaning. It is, unfortunately, the level at which our common schooling generally stops, and we read the language haltingly, in the manner described by J. V. Cunningham,

> With forward eye and finger on
> The grammar and the lexicon.

The poet was speaking of poetry. But whereas we are most of us ashamed that we do not read well, we know it is fashionable to assert our mathematical illiteracy.

The law of gravity becomes a *Gestalt*, a pattern of quality, in proportion as its implications and analogies are pursued and become familiar; and in the same proportion its "prosaic meaning" as an algebraic formula recedes into the background.[11] The perception of quality is the goal, but quantitative expressions provide the mode in which we can begin to grasp quality that lies beyond the range of customary experience.

So it is that mathematics has changed its position in the world of intellectual culture, from that of a tool of economic and technical interests, to that also of an indispensable tool of intellectual growth to be prized and to be advanced, and, therefore, for some, to be regarded as an end in itself.

The new use of quantitative techniques in science—to refine perception and thus extend its range—is only half the contribution of classical mathematics. The other half was the definition of problems on a new level. The Greeks developed the first rigorous and axiomatic exposition of the rational structure of ideas. Their demand for rigor meant that they also began to study the nature of theory as such and the techniques of constructing a theory. In so doing they extended the habits of the geometer to other spheres, and in formal logic they developed a critique of those habits, that is to say, a collection of standardized ways of testing whether these habits were consistent with each other.

The inspiration for the study of logic came fully as much from the social sphere as from the mathematical. The multitude of Greek tribes and city states, and later the political enfranchisement of a wide social class, the town-dwellers who often held a balance of power between the aristocracy and the merchant class, led to a tradition of public debate more potent than urban society had yet produced. The vivid pictures which Plato has given us of the professional educators of his day make obvious one of the motives that existed for the investigation and codification of logical forms. The dilemma by which one unseated an opponent in argument and the dilemma by which one proved the incommensurability of the side and diagonal of a square represented different uses of the same logical form, which logicians could name and investigate independently of either, or any, application. The development of logic reflected and strengthened the sense of the importance of rigor in demonstrating the connections of ideas.

The earliest achievements of science show two major limitations: they are confined to the static and to the isolated, whether in geometry, mechanics, biology, or astronomy. The relations described by geometry are accessible to investigation by elementary means only because of the nearly ideal rigidity of the solid state under ordinary conditions. Greek mechanics, a science growing out of the rigger's art, was limited to what is now called statics and used only the most rudimentary ideas of dynamics. Astronomy dealt with the motions of the planets, but these motions are, and were discovered by the Greeks to be, virtually periodic and unchanging. Aristotle tells us that circular motion is the best, because it is most nearly eternal. Biology and medicine deal with systems which we are apt to think of as unstable and complexly interacting; yet the Greeks showed that it is possible to make progress in these sciences by classifying unstable and transitional phenomena under relatively stable categories, and concentrating attention upon these categories.

Both "isolated" and "static" are concepts which in contemporary thought carry connotations of disparagement, and we are apt to underestimate the potentialities of the mode of thought which they characterize. The work of Aristotle and his followers, however, shows how vigorous a scientific and intellectual tradition can be developed within that mode of thought.

Self-Maintaining Processes: Aristotle's Natural Teleology

The ideas of Aristotelian natural science and philosophy have their origin in the concepts we use for discussing our own individual and social conduct. We refer to his "natural teleology." This is not the only occasion, in the history of science, when ideas lifted from one context of usage have given fruitful guidance in others. The Aristotelian scheme was in fact so useful that it became, in later generations, an ingrained habit. The early investigators and philosophers of our own period were perforce rebels against the habit. We who are their descendants can be more just in our appreciation of the teleological mode, especially since in our own most advanced science of physics there are remarkable parallels to it. We have gotten over Aristotle, and it is time we got over getting over him.

The value in Aristotle's teleology of nature does not reside in its apparent tendency to attribute "unconscious purposes" to nature, but in quite another aspect of the teleological habits of thought. This value can be emphasized in the most paradoxical way by saying that although Aristotelian philosophy is essentially "static," it conceives nature primarily in terms of processes; and while it conceives things as "isolated," it describes this isolation not as sheer aloneness, but as the functional interdependence of autonomous elements within a harmonious whole. The key conception is Aristotle's notion of a system of processes whose form or pattern is stationary, like the constancy of the candleflame or the living organism or the cyclic paths of planetary motion. Such patterns are conceived as self-maintaining, each system behaving according to its own intrinsic "nature," and not coerced except "accidentally" by external agencies. Each system may require from its environment certain necessaries, as food or fuel or material to work upon; and each may contribute to the means of existence of other neighboring systems. But this is no tightly knit speculative metaphysics, such as some of the Stoic writers later developed, in which every accident is conceived as an unknown essential to the cosmic plan. Aristotle insists upon the reality of the accidental, the casual jostlings and interactions of things in the pluralistic and somewhat loose-jointed world that man's working knowledge of things reveals to him.

From the point of view of modern scientific thought, the Aristotelian description of nature is neither wholly wrong nor wholly adequate. It is a necessary first approximation, from which later science must depart but without which it cannot progress. The origins of species are no problem until their identity and their stability have been recognized and fully credited. The dynamics of motion presupposes the analysis and measurement of forces and masses in contexts which are wholly static, or dynamic in only a rudimentary sense. Irreversible changes in planetary motions—such as those caused by tidal drag—are detectable only when their paths have first been described as if ideally stationary.

The teleological system of ideas, moreover, is not limited in its applications to processes which are stationary in a simple sense. It can deal with development, fluctuation, and degenerative change. The pattern of life is recurrent from generation to generation, but within a single cycle it is intensely dynamic and adaptive to a wide range of chance variation in the environment. So long as norms can be established for such processes they can be described in terms of "ends" or "final causes," and their quantitative departure from the norms treated as chance fluctuations without deep or lasting significance.

Here of course lies the essential weakness of the Aristotelian scheme, its incapacity to admit the relative character of the distinction between what is conceived as "essential" and what is conceived as "accidental." Historically, however, this weakness was a strength. It permitted that great intellectual synthesis in which man appeared as wholly man, yet wholly part of the world-order, with a rigidly fixed "human nature" constituting at once his scientific explanation and his guiding norm—a nature which, however, was not prescribed by barren theoretical requirements, but left open to be filled in by honest observation and analysis.

These same virtues and limitations appeared in the revival of Aristotelianism in the Middle Ages, brought to a culmination by Thomas Aquinas. Here the Aristotelian world-order was unified with the Christian dogma of revealed religion, profoundly contrasting the stationary patterns of natural processes with the dramatic and unidirectional character of the supernatural order. In the Aristotelian pluralism, emphasizing both the reality of the accidental and the impossibility of a science of it, Christian writers found place for the miraculous. In Aristotle's conception science can only deal with uniformity; it can deal best with the absolute uniformity which permits description within a rigid deductive framework, and less successfully, but still not badly, with the probable and prevailing uniformities with which biology, ethics, and politics must be content. The accidental *per se* cannot be generalized about; thus no science. Seen from the earth, the miraculous can only be described as accidental, improbable according to natural causes, but not as impossible.

From Teleology to Mechanism

In this way among others the basis was established for that long history of conflict named by one of the partisans of science "The warfare of science and theology," in which now one and then another line of truce was attempted ("the limitations of science"). On the other hand, the whole history of science since the end of the Renaissance has been an invasion of the domain of what had previously been regarded as closed to natural knowledge. This was the case with the imponderables of the seventeenth and early eighteenth centuries, which became accessible to measurement as heat and electricity within a hundred years, and with the organic chemical compounds of a later day which only gradually became subject to chemical synthesis and structural description. This invasion reached its first great culmination in Isaac Newton's System of the World which set forth the recurrent order of nature-in-the-large as a consequence of dynamical laws. Most of the astronomical facts upon which Newton relied had earlier been formulated within the Aristotelian mode of description. Copernicus and Kepler, who overthrew the older Ptolemaic astronomy, appear to have been more Aristotelian in philosophy than Ptolemy himself.[12] They conceived their work as setting forth the stable patterns of planetary motion as ultimates of their science. Newton derived these laws from the principles of mechanics, established first by terrestrial experiments more intimately related, in idea and in spirit, to the concerns of mercantile technology than to the moral and religious imagery within which Aristotelianism had long operated. Such concepts as force and mass are of meaning to a mind that consults hands as well as eyes; they are "operational" characters of physical systems. Newton combined the centrifugal force law, which can be verified in the terrestrial milieu by simple experiments, with the purely observational law of Kepler which relates the orbital velocities of the planets to their distance from the sun. The result was the law of gravitation, confirmed then by myriad reapplications to the details of the solar system and later by direct measurement of minute gravitational forces between masses small enough to be manipulated in the laboratory.

Since the concepts of mass and force are operational abstractions, their use in the description of nature at large suggests an extraordinary transformation of attitudes. In older frames of thought the operational manipulatory abstractions, like the quantitative ones, were condemned to a minor place within the philosophic scheme. They characterize the point of view of the artisan, not the philosopher; they are instrumental not ethical categories (yet Archimedes musing upon the simple principles of statics had declared he could move the earth). Newton's marriage of heaven and earth finally established the mundaneness of celestial laws. Pascal declared that man's dignity lies in thought; thought in the new mode was, however, no longer the contemplative delineation of man's place in the universe, it was inseparable from programs of engineering control.

If this account of new attitudes ended with the age of Newton, it would remain merely prophetic. The pace of great technological innovations, though accelerating, was still slow by comparison with the age that followed, and science was still the scholar of technology, not its master. But subsequent developments support the prophecy. Chemistry found the "sure path of science" in the eighteenth century, laying the foundations for the chemical industries of the nineteenth. The chemical transformation of matter extended man's control in one direction, while physics began to study what we now call the transformations of energy. This related particularly to the study of the principles of the steam engine, the first machine designed not to transmit, but to originate, mechanical motion. The concept of energy transformation and the law of conservation of energy provided a framework in which to unify the older theory of mechanics with the new discoveries in thermodynamics, in chemistry, and in electromagnetism. By the end of the nineteenth century one could justly revive, but now in a sober and even pedestrian spirit, the old Heraclitean principle that nothing is immutable in nature save its laws of change. The Aristotelian hierarchy of fixed natural kinds had been replaced by the conception of nature as a system of materials and energies plastic to human control.

The Aristotelian world-picture had been supplanted even earlier by the conception of a universe whose temporal changes were subject to general transformation laws, without reference to teleology. But it was only partially supplanted, and a good deal of intellectual history in the seventeenth and eighteenth centuries is unintelligible in terms of the usual account of the downfall of Aristotelianism in that period. To reject the ideas of earlier thought can only mean to reject what one can formulate and dispense with. It does not mean and cannot mean to reject what one is unconscious of having accepted.[13] A system of ideas is like an iceberg, mostly beneath the surface. The part of Aristotelianism above the surface in the Renaissance and seventeenth century was its mechanics, and this was rejected. But the laws of mechanics are very abstract and significant for prediction and control only when accompanied by a detailed account of the situations to which the laws apply. The mechanical laws of nature are consistent with an infinite variety of situations, on the cosmic scale of "possible worlds," and the question remains, why this one?

In the seventeenth century this question marked the boundary between science and theology, mechanism and teleology. On the other side of the boundary from science, teleological arguments still held sway, as we see in the philosophical writings of Leibniz and the correspondence of Newton. Nevertheless the boundary was an unstable one, and the deism of that age was possible only if certain questions were not pressed too hard. The theology which professed to find room for the Newtonian world-machine was already inadequate to the traditions of Christianity, and yet not really adequate to the needs of science. Even Newton's laws themselves, in their generality, point beyond the conception of a closed harmonious order to an open infinite world whose local regularities are constantly being disturbed and modified.

The wider implications of Newtonian mechanics were not pressed until much later when there was a readiness to examine them. In Newton's time, and in the greater part of the eighteenth century, the Aristotelian ideas of design and harmony held sway, and the laws of mechanics were conceived as serving to maintain a well-designed order inviolate through time.[14] It is in this period that Quesnay laid the foundations of scientific economics with his conception of an *ordre naturel* maintaining itself through a self-reproducing cycle analogous to the circulation of the blood. And Adam Smith's great work was aimed to comprehend the mechanisms of this self-maintenance.

From Mechanism to Evolution

The sense of "time's arrow" was not absent in the earlier period, but it was confined to special interests. Leibniz had conjectured that the earth was once a burning star, and had suggested the explanation of fossiliferous rocks. Kant noted that tidal drag implied a one-way evolution of the earth's orbit, and developed the evolutionary hypothesis later elaborated by Laplace as a grand outline of the evolution of the physical universe. For a century before Darwin the genetic explanation of the kinship of the varieties of life had been under discussion. Darwin contributed to the integral picture of life's history on the earth, but even more significantly he related this to a differential mechanism of evolution, the processes of variation and selection that could be studied experimentally in relation to the arts of plant and animal breeding.

With the Darwinian revolution, the growth of scientific geology, and the beginnings of modern archeology, the historical dimension became part of the framework of science, and the earlier boundary between science and teleology became more difficult to maintain. In Aristotle, in medieval philosophy, even in the seventeenth century, teleology was called upon to explain the appearance of adaptedness in nature. But this was only quasi-explanation, based upon the analogy of the work and the worker of art, appealing enough in an age of handicraft technology. The barrier collapsed with the first real progress in the art of asking evolutionary questions and answering them by empirical means. The old phrase "improbable according to natural causes" was then recognized as only a mask for unexamined beliefs about "natural causes." If the present state of an organism or a planet or the galactic systems is sufficiently improbable according to some hypothesis of its earlier history, so much the worse for that hypothesis. Conversely, the extraordinary adaptation we find about us in nature argues for nature a history equally extraordinary, by comparison with uncritical assumptions extended from common sense.

Law and Will: Science and Man's Image of His Own Nature

If the development of physics in the last three centuries implies a transformation of attitudes toward nature, the development of evolutionary science in the last century implies an equal alteration of attitudes toward man. Aristotelian man found himself a stranger in Newton's universe, albeit a purposeful and aggressive stranger. But Newton's universe was only an abstract, a single layer of the "world-onion." As new layers have been explored, there have indeed been further "alterations in the points of view best suited for the ordering of our experience," and man and nature have been joined together again, within the framework of evolutionary conception—not as Aristotle joined them, or the medieval church, or Newton, but joined nevertheless.

In seventeenth- and eighteenth-century materialism man is described as a perfected mechanism, *l'homme machine*, and the image is still that of levers and gears, of clockwork. Nineteenth-century materialism, in the style of such thinkers as Haeckel and Huxley, saw man as a product of biological history, incomplete and still evolving. But the process of evolution was still natural selection, the law of tide-pool and jungle. As such it lacked ethical relevance, although of course it had no lack of moralizing interpreters among the "social darwinists." The continuity of human development from pre-human ancestors, moreover, made easy the fallacy of denying man any status different from theirs and that of his animal cousins.

Only with the maturing of archeology and anthropology has man the object of scientific description come at all to resemble man the responsible moral agent. If lions once were gentle and doves fierce they changed unwittingly, by passive adaptation. Only man—to use Childe's phrase—"makes himself." That he does so is a result of biological evolution, but it is not biological evolution. In man the biological realm has evolved a new mode of evolution; its mechanism is not the law of the jungle, but knowledge and the freedom which knowledge bestows. In its own good time and by its own criteria of truth, science has found its version of the story of the expulsion from the garden of Eden.

Important as the evolutionary dimension of science has become, particularly for new information about, and new ways of conceiving, man's place in nature, there are developments in fields apparently remote from biology and anthropology that may prove equally indispensable for the further progress of science, and may involve equally radical revisions of its general conceptions of methodology. As one of the assumed barriers to systematic knowledge was the contrast between the objective teleology of nature and the purely mechanical transformation laws of physics, so another barrier has been between those same laws and the subjective teleology of human action (including that part of social evolution which must be described as deliberate achievement). It has long been recognized, indeed, that there is not so much an explicit contradiction as a radical discontinuity between the postulate of causality and the assumed reality of volition. By and large, the tradition of science has been to espouse the former and ignore or deny the latter. The special antagonistic reaction toward science from religion, philosophical idealism, and humanism has been caused more by this fact than by any other.

The scientist appears, in the usual picture of him, as a peculiarly object-minded individual, unwilling to accept as knowledge the self-conscious mode of experience which is stock-in-trade for the humanist, and upon which in the last analysis (so the idealist will insist) even science itself depends. All perception and all inquiry in the traditional unself-conscious mode of science involve the isolation of an object not only from its material environment, but also from the habits, memories, and desires of the percipient or investigator. Man is a part of the material world; but he views that world from a point of view which is not, then and there, perceived as a part of what is viewed. To a certain sort of critic of science the spectacle of the behaviorist in psychology who quite consciously (so to say) denies the reality of consciousness is humorous or, more often, to judge by the effect, extremely frustrating. Apart from the complicated semantic questions which such criticisms raise, there is undoubtedly this much validity in them: that in the whole history of modern science there has been a kind of unself-consciousness, or absorption in subject-matter, of naïveté, which it would be a mistake to suppose merely unwitting. Many scientists may be maladept at introspection, but that is not the real point. In the struggle against ignorance, the strategy of science is to divide and conquer. Having found the sure path in one small subject of investigation, and penetrated deeply into it, a good scientist seeks to go on from there into some other area that may seem at first unrelated. He mistrusts the division of experience into neat areas insulated from each other as though by legislative decree; the only boundaries he will acknowledge are those which the growth of knowledge itself discovers; and he is accustomed by experience to expect that those boundaries will be in some degree relative and fluid. Science may divide and conquer; but its faith is the unity of nature.

Science does not respect the barriers fixed by philosophy. The physiologist studies "perception," and makes naïve pictures of the process: light enters the eye, is focused on the retina, stimulates the optical end-organs, neural signals are transmitted to the brain, and perception supervenes. The philosopher points out to him that he studies not perception itself, but only some of its causes. There is nothing so absolute as the distinction between a thing and its causes; but whether from real or feigned naïveté, the physiologist refuses to be impressed, and continues to suppose that he is studying perception, although admittedly only from one special vantage-point. History would support him. The whole history of the theory of perception and knowledge in modern times shows the impact of discoveries in such fields as optics and acoustics. The tactile conception of visual perception, the casting of a glance, according to which the eye sends out something that touches a surface perceived, is native to pre-scientific common sense, yet it was set aside in view of physical and physiological discoveries. Contemporary theories go further. They insist that all perception implies the recall of patterns from memory. Scientific discovery is not the solution of philosophical questions, but it forces their reformulation.

In such connections as these, the humanist's revulsion against the *esprit simpliste* in science deserves a second look. F. E. L. Priestley refers to Swift's satires of the mechanistic theories of language that grew out of the Baconian tradition in which, as Swift and I think Priestley see it, there is a contempt for the arts. On further reflection, the subsequent story of Swift's language frame offers an instructive example. For it appears that the unnamed professor of the Grand Academy of Lagado began a sequence of researches which, today, bid fair to become one of those deep salients driven into previously unorganized subject-matter of knowledge, exhibiting the mode of progress characteristic of science. Satire and science have equal but opposite interests in perceiving the mechanical frame of human life. Swift's interest is to attack the narrowing and dehumanizing of life through preoccupation with the mechanical. The interest of science is to explore the mechanical (which it knows how to explore) to discover where its limits may be found, and of what kind they are. The recent researches in this field go under the heading of "Information Theory." Language can be regarded as a sequence of spoken or written symbols constructed by a corresponding sequence of choices from among the set of all those possible. These choices and combinations of choices can be characterized by the frequency of use by the total population and thus by a measurable probability. Empirical researches of cryptanalysts, communications engineers, and teachers of language carry on the work of Swift's professor, who "had made the strictest computation of the general proportion there is in books between the numbers of particles, nouns, and verbs, and other parts of speech."

Now one of the most interesting developments of recent times is the elucidation of the conception of the "entropy of information," by which the theory of communication is brought into relation with the physical concept of entropy as previously developed in statistical mechanics, and thus also with fundamental problems of physical and biological order. In none of this is there any implication that language is a random mechanical combination of elements. On the contrary, what characterizes the vital use of language is simply a very high thermodynamic improbability. Within the range of usage, moreover, there is a sliding scale. Advertising slogans and the numbered messages of the telegraph company represent one extreme, while certain kinds of poetry, the prose of Joyce, or the papers of Professor Einstein would represent the other. The latter are highly improbable combinations of symbols. The literary artist who is not sensitive to the statistical frequencies and associations of the symbols he uses is unable adequately to control his medium. The common word-associations may be the right ones for a particular purpose; or again, meaning may be enhanced by the deliberate use of highly improbable combinations, as when Emily Dickinson says of death that it is "at best an ablative estate."

The mathematical theory of the language frame will have its uses in technology, in the efficient coding, decoding, and transmission of information; and it will undoubtedly have other uses of this kind. But what is most important about it is that it provides a theoretical perspective within which communication is seen as an expression of the basic feature of life, that it is able to maintain itself at a higher level of energy and organization than its inorganic environment. The "entropy of information" is not an adequate conception of order in communication, and indeed entropy in general does not provide an adequate conception of even biological order; but it is a common denominator, and a beginning.

The greatest philosophical importance attaches to twentieth-century extensions in physics, especially the relativity theories of Einstein and the quantum theory. In both connections the "alterations of our points of view" needed to grasp these theories have been widely discussed, inside and outside the circle of physical scientists. Because of the very minor place which ideas even of classical science have had in contemporary culture, these remarkable discoveries, while attracting attention to the intellectual conquests of science, have also at the same time suffered a considerable vulgarization, contributing even to anti-intellectual tendencies essentially inimical to science. Both discoveries compel attention to the fact that the common sense description of nature is adequate only under conditions which common sense takes for granted but which are not, in fact, always satisfied. Relativity theory compels attention to the dependence of certain characteristics of physical phenomena upon the relative motion of an observer. Although this dependence is negligible in the range of velocities ordinarily encountered in experience, its recognition forces us to become self-conscious about ingrained ideas of space and time which prove inadequate in the range of higher velocities. In the case of quantum mechanics a similar revision is required. According to quantum theory there is an ultimate indivisibility and atomicity in all physical transitions, and this atomicity implies, in turn, that it is impossible in principle to achieve a perfectly sharp separation between "observation" and "reality."

Both these discoveries are of a sort to evoke a high degree of self-consciousness as to the use of basic conceptions underlying naïve "object-mindedness" in science. As a result there has arisen among physical scientists working in these fields a style of discourse and analysis which deliberately seeks to probe the limitations of customary conceptual conventions. This style has a strongly critical flavor, reminiscent of the anti-metaphysical traditions of western and oriental philosophy, the *via negativa* in theology, and the general attitude suggested by the maxim of Suarez: "Question the questions and answer the answers."[15]

Self-consciousness is always seized upon, by the hostile, as weakness. The negative side in modern physics is a loss of assurance in the unlimited adequacy of certain classical conceptions of science. Thus the popular "relativism" announced in Einstein's name, thus the "abandonment of causality" proclaimed by defenders of irrationalism on behalf of atomic physics. Such proclamations imply a failure to understand the essential conservatism of science in its progress, which abandons former points of view only when, and to the extent that, they are replaced by more adequate ones. The "relativism" of relativity is only the negative side of a positive advance, an extension of knowledge which would be as well and as badly described by the word absolutism as by relativism. In quantum mechanics, likewise, the "uncertainty principle" in its usual popular presentation is only a negative interpretation of a positive and well-verified law, which implies super-mechanical stabilities in atomic processes just as directly and immediately as it implies the absence of purely mechanical stability. Indeed, the two aspects are separable from each other only from the point of view that is already superseded.

The Participant Observer

The general importance of these discoveries of modern physics for science as a whole arises partly from their direct implications in cosmology and chemistry and biology, and partly because of the exemplary role which physics has played in modern times, as the most advanced science and thus in a sense the model for the other sciences. Any direct "imitation" of physics in other fields is of course a sign of immaturity; but if self-consciousness as to the participant role of the experimenter is part of a positive advance in physics, it may justifiably lead to revival of attention to similar problems in the biological and social sciences, ignored or suppressed under the influence of physics in an earlier stage. In particular it has really been obvious all along that the development of human individuals and societies is affected by self-knowledge. An unwillingness to receive facts of this order into the serious subject-matter of social science can at any rate no longer be justified by appeal to the model of physics.

Recognition by social scientists of the participant role of the observer has aspects with no parallel in physics, and presents a much richer series of problems. Thoughtful commentators have long realized that the description of human affairs, even though undertaken with the utmost honesty and freedom from prejudice, is essentially influenced by the particular nexus of practical relations which the observer has to those affairs, and which constitute, in an important sense, his means of observation. The existence of different schools of thought in political science, economics, and psychology can hardly be understood otherwise.

Accelerated Growth and Autonomy of Science

There has been little emphasis, in this essay, upon the detailed historical connections of science with other human affairs. But insofar as science has itself a direct dependence and concern with such connections, this adumbration is incomplete, and in conclusion something should be said about the matter. The connections are conveniently classified in two ways, as relating science to material aspects of culture on the one side, and to humanistic aspects on the other. On the material side the connection of science with technics is one of increasing strength. Science grows from technology, but grows at first away from it, and in its wake. Its freedom from routine and its progress in new directions, however, eventually reverse this tendency, and changes in technology are induced which would not have occurred spontaneously. New technology, once established, extends the range of hands and eyes, and the spiral continues. The Industrial Revolution was not so much the beginning of this process as its acceleration to a rate much faster than that of other determinants of social change.

This acceleration is important both for its social implications and for its impact on science. The coupling of science and technology has already revolutionized human existence, and we know that this revolution may be only a beginning. In our anxiety to resolve the problems which this revolution has posed and will continue to pose, we have too seldom considered the impact of these changes upon science itself. In particular the autonomy of science emphasized above—its freedom, the material and spiritual guarantees of support in pursuing its own characteristic aims—can hardly be unaffected by the intensified association with industrial and military innovations.

In many ways the intensification is a boon to science. As it has become more consequential in its impact, science has grown in prestige, in economic support, and in the power of inquiry. But while the power to inquire is not a power that corrupts, science as an institution can be corrupted. Society possesses a rich storehouse of means for identifying prestige with thoughtless respectability, and as scientists become more obviously important to the workaday world, they become readier targets for subtle combinations of threat and flattery. As the technological relations of science become more dynamic, its humanistic relations are threatened. And scientists are not as ready to meet this threat as they may have been in the past or must become in the future. The gap which rapid growth has left between science and the older humanities has injured science no less than humanism. The stereotype of the narrow lab man in the dirty apron may be a piece of aristocratic snobbery, but it includes a real criticism. The humanistic tradition, from Plato and Jesus, has long known that all special interests must be subordinated to the development of the whole, and that there is a pathology of vocation which is more destructive morally than some of the better advertised sins. When the physicist or economist is trained to offer his services to the highest, or even the only, bidder, and to avoid the public debate over ends on grounds of professional restriction, it is time to speak of a professional disease.

Science was born out of the critical spirit, and at its cutting edge it is criticism. In our time we have seen efforts, some of them of almost fantastic crudity, to partition the critical spirit, to put it in blinders. The result here too is one long known to the humanistic tradition. Science has its own resources to meet the problem in part; its pride of tradition, its method, and its results in the field of social study. Where science goes beyond criticism to creation it reaches again into a world of vast richness of possible new combinations and of great boldness in selecting from them the improbable but relevant—a world of creativity so rich and free that its thought is not easily broken to the yoke of external restrictions and controls. Here again the relevant traditions are not characterized by narrow separatism. From the time of Greece and the Middle Ages they imply the community of the humanistic spirit in its reliance upon the method of intelligence and its devotion to the common welfare. The need for a fuller attainment of this community may be allowed, today, as the most significant of the humanistic concerns of science.

NOTES

1. Herbert Butterfield, *The Origins of Modern Science, 1300–1800* (London: G. Bell, 1949). This book contains an excellent essay on the significance of science in modern history.

2. The novels of Hans Otto Storm are written out of the life of an engineer. One of them, *Pity the Tyrant* (New York: Longmans, Green & Company, 1937), republished in *Of Good Family* (New York: Swallow Press, 1948), is, or very nearly is, of first rank. All of Storm's novels are of first rank among their kind, limited as it is. The earliest is *Full Measure* (New York: The Macmillan Co., 1929), and the most ambitious is *Count Ten* (New York: Longmans, Green & Company, 1940).

3. *On Understanding Science* (New Haven: Yale University Press, 1947).

4. A recent able survey of evolution is that of G. G. Simpson, *The Meaning of Evolution* (New Haven: Yale University Press, 1949). On the relations of biological and social evolution and on human nature in archeological perspective, see the writings of V. Gordon Childe, especially *Man Makes Himself* (London: C. A. Watts & Co. Ltd., 1936).

5. Hans Otto Storm's reflections on this transition are contained in an essay, "Eolithism and Design," *Colorado Quarterly*, Summer, 1952.

6. Whatever his demerits, Oswald Spengler in the *Decline of the West* (New York: Alfred A. Knopf, 1926–8) has the great merit that he perceived the aspect of style in all human activities.

7. André Malraux, *Les Voix du silence* (Paris: Nouvelle Revue Française, 1951), p. 332.

8. The quotation is from the first essay in *Atomic Theory and the Description of Nature* (Cambridge: Cambridge University Press, 1934).

9. Benjamin Farrington is prone to do this. But his writings are full of substance for those who would pursue the relations of science and culture; see *Greek Science, Its Meaning for Us* (London: Penguin Books, 1944), *Science and Politics in the Ancient World* (London: Allen and Unwin, 1939), and *Head and Hand in Ancient Greece* (London: C. A. Watts & Co. Ltd., 1947).

10. The classical essay on the changes of quality with size is "On the Size of Things" in *Man Stands Alone* (New York: Harper Bros., 1941), by Julian Huxley.

11. "Même les proportions géométriques deviennent sentiments, car la raison rend les sentiments naturels et les sentiments naturels s'effacent par la raison." Blaise Pascal, *Pensées*, Louis Lafuma, ed. (No. 914; Paris: Jacques Delmas et Cie, 1952).

12. Butterfield shows how little truth there is in the notion of a sort of universal rejection of Aristotelianism.

13. The chapter in Butterfield, *The Origins of Modern Science*, on "The Conservatism of Copernicus" is illuminating in this respect.

14. Compare the chapter on "The Eighteenth Century and the Idea of Order," in J. Bronowski, *The Common Sense of Science* (Cambridge: Harvard University Press, 1953).

15. For one example, the essays of Bohr referred to above; for another, J. R. Oppenheimer's *Science and the Common Understanding* (New York: Simon and Schuster, 1954).

Lightning Source UK Ltd.
Milton Keynes UK
UKHW030612210722
406167UK00006B/700

9 781442 651388